Volume 28

CRM MONOGRAPH SERIES

Centre de Recherches Mathématiques
Montréal

Skew-Orthogonal Polynomials and Random Matrix Theory

Saugata Ghosh

The Centre de Recherches Mathématiques (CRM) of the Université de Montréal was created in 1968 to promote research in pure and applied mathematics and related disciplines. Among its activities are special theme years, summer schools, workshops, postdoctoral programs, and publishing. The CRM is supported by the Université de Montréal, the Province of Québec (FQRNT), and the Natural Sciences and Engineering Research Council of Canada. It is affiliated with the Institut des Sciences Mathématiques (ISM) of Montréal. The CRM may be reached on the Web at www.crm.math.ca.

American Mathematical Society
Providence, Rhode Island USA

The production of this volume was supported in part by the Fonds Québécois de la Recherche sur la Nature et les Technologies (FQRNT) and the Natural Sciences and Engineering Research Council of Canada (NSERC).

2000 *Mathematics Subject Classification.* Primary 33-XX, 11Cxx, 26Cxx, 15–XX.

For additional information and updates on this book, visit
www.ams.org/bookpages/crmm-28

Library of Congress Cataloging-in-Publication Data

Ghosh, Saugata, 1970-
Skew-orthogonal polynomials and random matrix theory / Saugata Ghosh.
p. cm. – (CRM monograph series ; v. 28)
Includes bibliographical references.
ISBN 978-0-8218-4878-4 (alk. paper)
1. Orthogonal polynomials. 2. Random matrices. I. Title.

QA404.5.G48 2009
515'.55–dc22

2009029006

♾ The paper used in this book is acid-free and falls within the guidelines
established to ensure permanence and durability.
This volume was submitted to the American Mathematical Society
in camera ready form by the Centre de Recherches Mathématiques.
Visit the AMS home page at `http://www.ams.org/`

10 9 8 7 6 5 4 3 2 1 14 13 12 11 10 09

. . . To my family

Contents

Acknowledgments

A major part of the work was supported by The Abdus Salam International Centre for Theoretical Physics, Trieste, Italy. Prof. Virgil Pierce helped me in locating some references and offered useful suggestions which contributed to the final presentation of the manuscript. Thanks to Prof. Arthur Greenspoon (Mathematical Reviews) and André Montpetit for their careful proof reading. I am grateful to Profs. M. L. Mehta, John Harnad, and Bertrand Eynard for their encouragement and help.

CHAPTER 1

Introduction

Orthogonal, bi-orthogonal and skew-orthogonal polynomials appear in the study of statistical properties of random matrix models. For example, 1-matrix unitary ensembles of random matrices (UE) require orthogonal polynomials $P_n(x)$ of order n, defined as

$$\int_{\mathbb{R}} P_n(x)P_m(x)w(x)\ \mathrm{d}x = h_n\delta_{n,m} \qquad n,m \in \mathbb{N}, \tag{1.1}$$

where $w(x)$ is the weight function and h_n the normalization constant.

Similarly, orthogonal ($\beta = 1$) and symplectic ($\beta = 4$) ensembles (OE and SE respectively) require an understanding of skew-orthogonal polynomials $\pi_n^{(\beta)}(x)$ of order n, defined as [27]

$$\begin{aligned}
&\int_{\mathbb{R}}\int_{\mathbb{R}} \pi_n^{(1)}(x)\pi_m^{(1)}(y)\epsilon(x-y)w(y)w(x)\ \mathrm{d}y\ \mathrm{d}x = g_n^{(1)}Z_{n,m}, \qquad &\epsilon(r) = \frac{|r|}{2r},\\
&\int_{\mathbb{R}} [\pi_n^{(4)}(x)\pi_m^{(4)\prime}(x) - \pi_m^{(4)}(x)\pi_n^{(4)\prime}(x)]w^2(x)\ \mathrm{d}x = g_n^{(4)}Z_{n,m}, \qquad &(') \equiv \frac{\mathrm{d}}{\mathrm{d}x},
\end{aligned} \tag{1.2}$$

where $g_n^{(\beta)}$ is a skew-normalization constant and Z is the canonical antisymmetric matrix with elements

$$Z = \begin{pmatrix} 0 & +1 & 0 & 0 & \dots \\ -1 & 0 & 0 & 0 & \dots \\ 0 & 0 & 0 & +1 & \dots \\ 0 & 0 & -1 & 0 & \dots \\ \dots & \dots & \dots & \dots & \dots \\ \dots & \dots & \dots & \dots & \dots \end{pmatrix} \equiv \begin{pmatrix} 0 & +1 \\ -1 & 0 \end{pmatrix} \dot{+} \begin{pmatrix} 0 & +1 \\ -1 & 0 \end{pmatrix} \dot{+} \cdots \tag{1.3}$$

with 2×2 blocks

$$\begin{pmatrix} 0 & +1 \\ -1 & 0 \end{pmatrix} \tag{1.4}$$

along the leading diagonal while the rest of the terms are zero. The antisymmetric integral in (1.2) is nonzero for each pair of polynomials $\pi_{2l}^{(\beta)}(x), \pi_{2l+1}^{(\beta)}(x)$, implying thereby separate patterns for even and odd order polynomials.

For unitary 2-matrix models, we require an understanding of bi-orthogonal polynomials $Q_n(x)$ and $\Pi_n(x)$ of order n, defined by

$$\int_{\mathbb{R}}\int_{\mathbb{R}} Q_n(x)\Pi_m(y)w(x,y)\ \mathrm{d}x\ \mathrm{d}y = h_n^{(\mathrm{BO})}\delta_{n,m}, \tag{1.5}$$

where $w(x,y)$ is the bivariate weight function and $h_n^{(\mathrm{BO})}$ the normalization constant.

The rich literature available on orthogonal polynomials [10, 18, 70, 77] and references therein, and some recent developments in the study of bi-orthogonal polynomials [6, 8, 9, 46, 50, 53] have contributed to our understanding of the unitary ensembles. In this book, we aim to develop the theory of skew-orthogonal polynomials and apply them in random matrix analysis, so that the three ensembles arising from the three-fold symmetry of Dyson can stand on an equal footing.

We have considered skew-orthogonal polynomials corresponding to generalized weight functions, defined in (3.23) – (3.25). For applications in random matrix theory, we have restricted ourselves to the simplest cases of the above weight functions, namely classical weights and the first nontrivial polynomial potential, i.e., weight functions with a quartic double well potential.

By classical weights, we mean the Jacobi family of weights (namely Jacobi, Gegenbauer, Chebyshev, Legendre weights), including the limiting cases of Gaussian and associated Laguerre weight functions. The Jacobi weight function is defined in the interval $[-1, 1]$ by

$$w_{a,b}(x) = (1-x)^a(1+x)^b, \qquad a, b > -1. \tag{1.6}$$

The associated Laguerre weight function is defined in the interval $[0, \infty]$ as

$$w_a(x) = x^a \mathrm{e}^{-x}, \qquad a > -1, \tag{1.7}$$

while the Gaussian weight function is defined in the interval $[-\infty, \infty]$ as

$$w_{\mathrm{G}}(x) = \exp(-\beta x^2/2). \tag{1.8}$$

Restrictions on the parameters "a," "b" for Jacobi and "a" for associated Laguerre ensures that these weight functions have finite moments (2.15). By "quartic weight," we mean

$$w(x) = \exp[-NV(x)], \qquad \text{where } V(x) = \frac{gx^4}{4} + \frac{tx^2}{2}, \; g > 0, \; t < t_{\mathrm{c}}, \tag{1.9}$$

where t_{c} is the critical value of t where we see the birth of a cut. Here, we note that our methods are independent of the parameters of the weight functions and hence can be used to study a much wider class of random matrix ensembles.

At this point, one is tempted to think that skew-orthogonal polynomials, like orthogonal polynomials, will find applications in other branches of physics and mathematics [48, 49, 60]. For example, there are differential equations (integrable systems), combinatoric, geometric and representation theoretic reasons one might wish to study skew-orthogonal polynomials directly. However even from random matrix theory there is a reason to consider these. Universality and the technique of controlling the error between the kernel functions comprised of orthogonal and skew-orthogonal polynomials is only going to get us to the leading orders of the resulting expressions, and that too via some cumbersome calculations. Once we begin to ask questions about higher order effects (which is a question that has bearing on the subjects we mentioned above) it will potentially be more efficient to have studied the skew-orthogonal polynomials directly.

1. 1-Matrix model: Orthogonal and skew-orthogonal polynomials

Random matrix ensembles follow the three-fold classification [24] to take into account the fundamental symmetries of the system. However, in such matrix ensembles, the condition of "statistical independence," is assumed somewhat artificially for the Gaussian ensembles. This leads to the study of non-Gaussian ensembles of $N \times N$ Hermitian matrices H with probability distribution

$$P_{\beta,N}(H)\,\mathrm{d}H = \frac{1}{Z_{\beta,N}} \exp\bigl(-\operatorname{tr} u(H)\bigr)\,\mathrm{d}H, \qquad \beta = 1,2,4, \tag{1.10}$$

$\mathrm{d}H$ is the Lebesgue measure. The matrix function $u(H)$ is (i) a polynomial in H or (ii) defined by the power expansion of the function $u(z)$. $Z_{\beta,N}$ is the so-called partition function:

$$Z_{\beta,N} = \int_{H\in M_N^{(\beta)}} \exp\bigl(-\operatorname{tr} u(H)\bigr)\,\mathrm{d}H = \begin{cases} N!\prod_{j=0}^{N-1} h_j, & \beta = 2,\\ N!\prod_{j=0}^{N-1} g_j^{(\beta)}, & \beta = 1,4, \end{cases} \tag{1.11}$$

where $M_N^{(\beta)}$ is the set of all $N \times N$ Hermitian ($\beta = 2$), real symmetric ($\beta = 1$) and quaternion self-dual ($\beta = 4$) matrices. h_j and $g_j^{(\beta)}$ are normalization constants for monic orthogonal and the two families of monic skew-orthogonal polynomials respectively.

From (1.10), using the invariance properties, one can obtain the joint-probability distribution of the N-eigenvalues $(x_1, x_2, \ldots, x_N) \in \mathbb{R}^N$:

$$P_{\beta,N}(x_1,\ldots,x_N) = c_{\beta,N}|\Delta(x_1,\ldots,x_N)|^\beta \prod_{j=1}^{N} w(x_j), \qquad \beta = 1,2,4, \tag{1.12}$$

where $c_{\beta,N}$ is a normalization constant.

$$\Delta(x_1,\ldots,x_N) = \prod_{j<k}(x_j - x_k) \tag{1.13}$$

is the Vandermonde determinant and comes from the Jacobian of the transformation.

Fluctuations in the eigenvalue spectra can be studied by the n-eigenvalue or n-level correlation function [27,53] $R_n^{(\beta)}(x_1,\ldots,x_n)$ which is defined by

$$R_n^{(\beta)}(x_1,\ldots,x_n) = \frac{N!}{(N-n)!}\int_{\mathbb{R}^{N-n}} \mathrm{d}x_{n+1}\cdots \mathrm{d}x_N P_{\beta,N}(x_1,\ldots,x_N). \tag{1.14}$$

This gives the probability density of finding n eigenvalues at $x_1, x_2, \ldots, x_n$, irrespective of the positions of the remaining eigenvalues $(x_{n+1}, x_{n+2}, \ldots, x_N)$, which are hence integrated out. $R_1^{(\beta)}(x)$ is the level density, with

$$D^{(\beta)}(x) = \bigl(R_1^{(\beta)}(x)\bigr)^{-1} \tag{1.15}$$

being the mean level spacing at x for large N.

In the ensembles we consider in this book, $R_n^{(\beta)}$ is calculated in the bulk of the spectrum.

To study the n-point correlation function, (i.e., to evaluate integrals of the form (1.14)), Dyson and Mehta have shown that if one can

- write the joint probability distribution $P_{\beta,N}(x_1, x_2, \ldots, x_N)$ in terms of ordinary/quaternion determinants,
- satisfying certain normalization
- and convolution conditions,

then the integral in (1.14) can be calculated by removing the $N - n$ rows and columns of elements/quaternion elements from the above-mentioned determinant/quaternion determinants. This defines uniquely the role played by orthogonal and skew-orthogonal polynomials. For example, for $\beta = 2$, the first condition gives the 2-point correlation function:

$$R_2^{(2)}(x,y) = \det \begin{pmatrix} S_N^{(2)}(x,x) & S_N^{(2)}(x,y) \\ S_N^{(2)}(y,x) & S_N^{(2)}(y,y) \end{pmatrix}. \tag{1.16}$$

The level density is given by

$$R_1^{(2)}(x) := S_N^{(2)}(x,x), \tag{1.17}$$

where

$$\begin{aligned} S_N^{(2)}(x,y) &:= \sum_{n=0}^{N-1} \phi_n^{(2)}(x)\phi_n^{(2)}(y), \qquad \phi_n^{(2)}(x) = \sqrt{\frac{w(x)}{h_n}} P_n(x), \\ P_n(x) &= \sum_{j=0}^{n} k_j^{(n)} x^j. \end{aligned} \tag{1.18}$$

Normalization and convolution properties ensure that $P_j(x)$ are orthogonal polynomials satisfying (1.1). Once the polynomials are defined, the sum in (1.18) is known [77] and is called the Christoffel – Darboux formula.

For $\beta = 1$ and 4, Dyson and Mehta have shown that one can write the joint probability distribution $P_{\beta,N}(x_1, x_2, \ldots, x_{2N})$ of an ensemble of $2N \times 2N$ random matrices in terms of quaternion determinants (i.e., the determinant of a matrix, each of whose elements is a 2×2 quaternion) satisfying similar normalization and convolution properties [51]. Using the Dyson – Mehta theorem (see Chapter 2), one can calculate (1.14). For example one can evaluate the 2-point correlation function in terms of a quaternion determinant (to be defined in Chapter 2). The correlation functions for a $2N \times 2N$ ensemble of random matrices, invariant under orthogonal and symplectic transformations, are given by

$$R_2^{(\beta)}(x_1,x_2) = Q\det\left[Q_{2N}^{(\beta)}(x_j,x_k)\right]_{j,k=1,2}, \tag{1.19}$$

where each quaternion element is given in its 2×2 representation by

$$Q_{2N}^{(\beta)}(x,y) = \begin{pmatrix} S_{2N}^{(\beta)}(x,y) & D_{2N}^{(\beta)}(x,y) \\ I_{2N}^{(\beta)}(x,y) - \delta_{1,\beta}\epsilon(x-y) & S_{2N}^{(\beta)}(y,x) \end{pmatrix} \tag{1.20}$$

and the level density

$$R_1^{(\beta)}(x) := S_{2N}^{(\beta)}(x,x), \qquad \beta = 1, 4, \tag{1.21}$$

where

$$\begin{aligned} S_{2N}^{(\beta)}(x,y) &:= \sum_{j,k=0}^{2N-1} Z_{j,k}\phi_j^{(\beta)}(x)\psi_k^{(\beta)}(y), \\ D_{2N}^{(\beta)}(x,y) &:= -\sum_{j,k=0}^{2N-1} Z_{j,k}\phi_j^{(\beta)}(x)\phi_k^{(\beta)}(y), \\ I_{2N}^{(\beta)}(x,y) &:= \sum_{j,k=0}^{2N-1} Z_{j,k}\psi_j^{(\beta)}(x)\psi_k^{(\beta)}(y), \qquad S_{2N}^{(\beta)}(y,x) = S^{\dagger}{}_{2N}^{(\beta)}(x,y), \end{aligned} \tag{1.22}$$

and

$$\phi_n^{(\beta)}(x) = \frac{1}{\sqrt{g_n^{(\beta)}}}\pi_n^{(\beta)}(x)w(x), \quad \pi_n^{(\beta)}(x) = \sum_{k=0}^{n} c_k^{(n,\beta)} x^k, \qquad \beta = 1, 4, \tag{1.23}$$

$$\psi_n^{(4)}(x) = \frac{\mathrm{d}}{\mathrm{d}x}\phi_n^{(4)}(x), \qquad \psi_n^{(1)}(x) = \int_{\mathbb{R}} \epsilon(x-y)\phi_n^{(1)}(y)\,\mathrm{d}y. \tag{1.24}$$

Normalization and convolution properties ensure that these functions satisfy skew-orthonormality conditions (1.2). For skew-orthogonal functions $\phi_n^{(\beta)}(x)$ and $\psi_n^{(\beta)}(x)$ corresponding to generalized weight functions (3.23) – (3.25), we calculate the sum $S_{2N}^{(\beta)}(x,y)$ in (1.22) for $\beta = 1$ and 4. We call them the generalized Christoffel–Darboux formula.

Exercise.
- Prove that the skew-normalization constants satisfy the property $g_{2m}^{(\beta)} = g_{2m+1}^{(\beta)}$.
- Prove that apart from the normalization, the even-order polynomials are unique, but to the odd-order ones, any multiple of the *next* lower even-order polynomial can be added.

For the applications in random matrix theory, we choose a $d = 1$ case from each of equations (3.23) – (3.25) (the classical weights), as well as a $d = 2$ case of (3.25), namely the quartic double well potential. In this regard, one can even extend the study of skew-orthogonal polynomials to incorporate the idea of Balian where one may choose the global properties at will [3]. This essentially brings us to different weight functions for which one first needs to verify the existence of the skew-orthogonal polynomials. Having done that, one has to develop the theory of skew-orthogonal polynomials further until one can use them to study the corresponding random matrix ensembles.

However, the local properties, like the energy level fluctuations in the bulk of the spectrum for large random matrices, after suitable rescaling (termed in the literature as "unfolding")

$$x_j = x + r_j D^{(\beta)}(x), \qquad j = 1, 2, \ldots, n, \tag{1.25}$$

are universal. The universality result, different for the three symmetry classes, can be expressed in terms of the sine-kernel, viz.

$$\frac{1}{\pi}\frac{\sin \pi(x-y)}{(x-y)}. \tag{1.26}$$

Here, we note that the word "unfolding" means choosing the point x as the new origin and rescaling so that the level density at this point is equal to 1.

2. Universality

In this section, we discuss briefly the historical background of "universality" in physical systems and their relations with random matrix theory. For an updated account of the recent developments in this remarkable phenomenon, readers are referred to [19], the four books [20, 29, 42, 47] and the two long review articles [23, 43].

In the random matrix analysis of spectral correlations in heavy nuclei, the spectra of 30 nuclei were considered [13, 45]. The energy levels in each spectrum belonged to the same angular momentum and parity. The spacing distribution and number variance were obtained for the individual spectra as well as for the combined spectra (called the nuclear data ensemble (NDE)). The results showed remarkably close agreement with the GOE predictions.

This was followed by two major developments where it was shown that random matrix theory has a much wider range of applicability. The first is the discovery that the Gaussian ensembles of random matrices can be used to study signatures of quantum chaos in Hamiltonian systems [11, 12]. Numerical analysis of several quantum systems with chaotic classical counterparts (such as the Sinai billiard, kicked rotors, etc.) have shown that as long as the underlying classical system is chaotic, the quantum energy spectra display fluctuations which are characteristics of the Gaussian ensembles of random matrices. The results for the spacing distribution, for example, are observed to be the same as that obtained for the nuclear spectrum.

The second is the discovery of a relation between fluctuation properties of random matrices and the universal conductance fluctuations observed in disordered mesoscopic systems [4, 57]. This leads to the development of the random matrix theory of quantum transport.

In the context of the universality of the energy-level fluctuations, the study of billiards is of particular interest. A billiard is a dynamical system produced by the free motion of a point particle in a given domain bounded by a closed curve with elastic reflections at the boundary. Among the regular billiards (viz. billiards with integrable Hamiltonians), the rectangle and the circle are common examples. Among the chaotic billiards, the Sinai billiard and the stadium billiard are common examples. The Sinai billiard consists of the study of the motion of a particle in a square, with a circular obstacle at the center. In the stadium billiard, one studies the motion of a particle in a stadium shaped curve (i.e., two half circles, joined by two segments parallel to the line passing through their centers). If one studies the solutions of the Schrödinger equation [11] with the billiard as the boundary, one observes Poisson-like distribution for the regular billiards while the GOE distribution for the chaotic ones. For example, if one studies the solution of the Schrödinger equation with Sinai billiard as the boundary and with the symmetry

$$\psi(x,y) = -\psi(-x,y) = -\psi(x,-y) = -\psi(y,x) \tag{1.27}$$

one obtains the GOE type spacing distribution. Finally, if the spectrum is mixed, i.e., it contains levels from all the four symmetries, the fluctuations observed are intermediate between GOE and Poisson. Similar results are also observed in nuclear spectra if one is not careful in analyzing data corresponding to a given quantum number.

Similar analysis can also be made for nonconservative time-periodic Hamiltonian systems, also called the Floquet systems. It is appropriate to study such systems in terms of circular ensembles of random matrices. Like the Gaussian ensembles, corresponding to the conservative systems, the n-point correlation function and hence the spacing distribution for such systems show the same universal pattern. A typical example of such a system which has been studied extensively, both classically and quantum mechanically, is the kicked rotator [73] where the Hamiltonian is given by

$$H(t) = \frac{L^2}{2I} + k\cos\theta \sum_n \delta(t - n\tau). \tag{1.28}$$

The first term describes the rotation of a pendulum with angular momentum L, and moment of inertia I. The second term describes periodic kicks with period τ with a pulsed gravitational potential of strength $k = mgh$. The equations of motion can be obtained from Hamilton's equations. As L changes discontinuously, after the nth kick, one observes the well-known "standard map," given by

$$\theta_{n+1} = \theta_n + L_{n+1}, \tag{1.29}$$

$$L_{n+1} = L_n + k \sin \theta_n. \tag{1.30}$$

where θ_n and L_n are the variables just after the nth kick. It is a particularly good example to study the transition of a Hamiltonian system from regular to chaotic behavior when changing the control parameter k. For small k, the rotator behaves regularly for most initial conditions. For large k, the motion in phase space becomes diffusive, thereby demonstrating a transition from integrability to chaos.

Now we come to the study of such systems quantum mechanically. Since the Hamiltonian is time-dependent, it is not possible to separate the time dependence from the Schrödinger equation, with the consequence that the energy is no longer conserved. In such systems, the time evolution of any initial state vector ψ_0 is obtained from

$$\psi_n = F^n \psi_0, \tag{1.31}$$

where ψ_n is the state vector after the nth kick. F is called the Floquet operator and for the kicked rotator

$$F = \exp\left(-\frac{\mathrm{i}}{\hbar} k \cos\theta\right) \exp\left(\frac{\mathrm{i}\tau}{\hbar}\frac{L^2}{2I}\right), \tag{1.32}$$

where L is the angular momentum operator $-\mathrm{i}\hbar\partial/\partial\theta$. The matrix elements of F are easily determined in the basis of the eigenfunctions of L. Using finite-dimensional models of (1.32) one can study the fluctuation properties of the eigenangles of F. One finds that, for $k = 0$, the fluctuations are Poisson-like while for large k they are GOE-like.

Apart from these examples, universality is also observed in other systems. The spectra of the atoms, molecules and the distribution of zeros of the Riemann zeta function also show the fluctuations of the Gaussian ensembles.

3. 2-Matrix model: Bi-orthogonal polynomials

In this section, we give a brief introduction to the 2-matrix model and bi-orthogonal polynomials. This is done with the belief that skew-orthogonal polynomials are in some way related to the bi-orthogonal polynomials. However, we have no proof to justify our claim.

We consider two $N \times N$ Hermitian matrices with joint-probability distribution

$$P(H_1, H_2) \propto \exp\{-\operatorname{tr}\big(\tfrac{1}{2}[V_1(H_1) + V_2(H_2) - cH_1H_2]\big)\} \tag{1.33}$$

where

$$V_1(H_1) = \sum \frac{u_k H_1^k}{k}, \qquad V_2(H_2) = \sum \frac{v_k H_2^k}{k} \tag{1.34}$$

are even ordered polynomials with positive leading coefficient. The coupling constant c is such that the joint-probability distribution function (j.p.d.f.) can be normalized.

The j.p.d.f. of the eigenvalues of the matrices H_1 and H_2 are given by

$$\begin{aligned}(1.35)\quad & F(x_1,\dots,x_N;y_1,\dots,y_N)\\ &= C\exp\left\{-\frac{1}{2}\sum_{i=1}^{N}[V_1(x_i)+V_2(y_i)-cx_iy_i]\right\}\prod_{1\le i<j\le N}(x_i-x_j)(y_i-y_j)\\ &= C\prod_{1\le i<j\le N}(x_i-x_j)(y_i-y_j)w(x_i,y_j).\end{aligned}$$

We are interested to study the correlations that exist in the above-mentioned two-matrix model. For this, we define the correlation function

$$\begin{aligned}(1.36)\quad & R_{k,p}(x_1,\dots,x_k;y_1,\dots,y_p)\\ &:= \left(\frac{N!}{(N-k)!}\right)\left(\frac{N!}{(N-p)!}\right)\int\cdots\int F(x_1,\dots,x_N;y_1,\dots,y_N)\\ &\qquad\qquad dx_{k+1}\cdots dx_N\, dy_{p+1}\cdots dy_N,\\ &= \det[S(x_i,y_j)]\end{aligned}$$

where

$$S(x,y) := \sum_{l=0}^{N-1}\frac{1}{h_l}\psi_l(x)\phi_l(y). \tag{1.37}$$

Here,

$$\begin{aligned}\psi_l(x) &:= \int Q_l(y)\exp\left[-\tfrac{1}{2}(V_2(y)-cxy)\right]\,dy;\\ \phi_l(x) &:= \Pi_l(x)\exp\left[-\tfrac{1}{2}(V_1(x))\right],\end{aligned}\tag{1.38}$$

satisfying the orthogonality relation

$$\int \psi_l(x)\phi_k(x)\,dx = h_l^{(BO)}\delta_{l,k}. \tag{1.39}$$

4. Summary and preview

Orthogonal polynomials satisfy a three-term recursion relation irrespective of the weight function with respect to which they are defined. This gives a simple formula for the kernel function $S_N^{(2)}(x,y)$, known in the literature as the Christoffel–Darboux sum. Availability of asymptotic results of orthogonal polynomials and the simple structure of the Christoffel–Darboux sum makes the study of unitary ensembles of random matrices relatively straightforward.

In this book, we develop the theory of skew-orthogonal polynomials. We obtain recursion relations, which, unlike orthogonal polynomials, depend on the weight functions. Using this, we derive reduced formulas for $S_N^{(\beta)}(x,y)$ for $\beta=1$ and 4 for finite N. We call them the generalized Christoffel–Darboux formula.

To make an asymptotic analysis, we obtain a system of differential-difference-deformation equations satisfied by a finite sequence of skew-orthogonal functions. We wished to make a Riemann–Hilbert analysis of the "fundamental solutions" as done by Deift et al. for orthogonal polynomials (see [18] and references therein). We failed. This forced us to look for an alternate method, namely proving the existence of simple mapping relations between skew-orthogonal and orthogonal functions. Having obtained that, known asymptotic properties of orthogonal polynomials are used to study skew-orthogonal polynomials. Mehta and Mahoux [54] made a similar effort while studying partition functions with quartic weight, while Adler et al. [2], Nagao and Waditi [62, 63] studied correlation functions for classical weights. However, to get compact relations (and not a series) which are easily amenable to asymptotic analysis of the skew-orthogonal functions, one needs to make a very careful choice of the basis of expansion.

Using these asymptotic results of skew-orthogonal functions in the generalized Christoffel – Darboux formula, we study the above mentioned kernel function and hence the level density and correlation function for large N. After suitable local rescaling of the spectrum we show that the n-level correlation functions for $n > 1$ are universal, being independent of the weight function (hence the level density) and (in the bulk) independent of the location in the spectrum. Of course the universal correlation functions are of three types, one for each β value mentioned above.

In Chapter 2, we review the Dyson – Mehta theorem on quaternion determinants relevant in the evaluation of certain multiple integrals. These are needed in calculating the level densities and correlation functions of random matrix ensembles. At the end, we present an alternate (resolvent) method to obtain the level densities for large N, which are later verified by the polynomial method. This procedure, although successful, lacks mathematical rigor and is given only for fun.

In Chapter 3, we obtain the kernel $S_N^{(\beta)}(x, y)$, $\beta = 1, 2, 4$ and finite N. The Christoffel – Darboux sum for $\beta = 2$ is rederived using an alternate method and can be considered as a warm up to the slightly more complicated derivation of the generalized Christoffel – Darboux formula for $\beta = 1$ and 4.

In Chapter 4, we derive mapping relations between skew-orthogonal and orthogonal functions of order n, where n is finite. Using them in Chapters 5, 6, 7, and 8, we work out explicitly the level densities and correlation functions of the entire family of Jacobi ensembles (including the limiting cases of associated Laguerre and Gaussian ensembles) and quartic two-cut potential, for UE, OE (even dimensions), OE (odd dimensions) and SE respectively.

In Chapter 9 we show that a finite sequence of these skew-orthogonal functions satisfy a mutually compatible system of differential-difference-deformation equations.

Chapter 10 gives the matrix-integral representations of the skew-orthogonal polynomials for arbitrary weight in terms of averages over matrix ensembles.

Through the mapping relation, we also observe the existence of a dual relation between the two families of skew-orthogonal functions arising in the study of orthogonal and symplectic ensembles of random matrices. This is discussed in Chapter 11. The results are summarized in the concluding section.

CHAPTER 2

Level Density and Correlation Functions

Here we review the Dyson – Mehta theorem [26] (also see [52]) and the emergence of orthogonal and skew-orthogonal polynomials, while computing multiple integrals of $|\Delta(x_1, \ldots, x_N)|^\beta \prod_{i=1}^N w(x_i)$ on $\mathbb{R}^{N-n}$ for $\beta = 1, 2$ and 4.

In Section 1 we state the main theorem needed to evaluate such multiple integrals. The formal polynomial results for the integral is given in Section 2. In Section 3, we state the "universal" $n > 1$ results as $N \to \infty$ for the Gaussian ensembles. In Section 4, we give a nonpolynomial method for determining the level density ($n = 1$). Readers who wish to remain focused on the polynomial method can skip this section since we will derive the same results in later chapters using the polynomial method.

1. Dyson – Mehta theorem on correlation functions

We start this section with definitions of quaternion matrices and quaternion determinants along with a useful theorem relating quaternion determinants and ordinary determinants [52]. This is followed by the theorem required to calculate the integral (1.14) (i.e., the n-point correlation function) for the three ensembles of random matrix theory.

A quaternion is a linear combination of four units, $1, e_1, e_2$, and e_3 such that any quaternion q can be written as

$$q = q^{(0)}\mathbf{1} + \vec{q} \cdot \vec{e} = q^{(0)}\mathbf{1} + q^{(1)}e_1 + q^{(2)}e_2 + q^{(3)}e_3. \tag{2.1}$$

Here $q^{(0)}, q^{(1)}, q^{(2)}, q^{(3)}$ are real or complex numbers, while the four units satisfy

$$\mathbf{1}e_i = e_i\mathbf{1}, \qquad e_ie_j = -\delta_{ij}\mathbf{1} + \epsilon_{ijk}e_k, \quad i,j,k = 1,2,3, \tag{2.2}$$

where δ_{ij} is the Kronecker delta and ϵ_{ijk} is the antisymmetric tensor with $\epsilon_{123} = +1$. The scalar part of q is $q^{(0)}$. The dual of a quaternion is defined as

$$q^{\mathrm{D}} := -Zq^{\mathrm{t}}Z = q^{(0)}\mathbf{1} - \vec{q} \cdot \vec{e}. \tag{2.3}$$

Any quaternion can be represented by a 2×2 matrix with complex elements: for example

$$\mathbf{1} = \begin{bmatrix} 1 & 0 \\ 0 & 1 \end{bmatrix}, \qquad e_1 = \begin{bmatrix} 0 & -1 \\ 1 & 0 \end{bmatrix}, \qquad e_2 = \begin{bmatrix} 0 & -\mathrm{i} \\ -\mathrm{i} & 0 \end{bmatrix}, \qquad e_3 = \begin{bmatrix} \mathrm{i} & 0 \\ 0 & -\mathrm{i} \end{bmatrix}. \tag{2.4}$$

Similarly, any 2×2 matrix with complex elements can be represented by a quaternion.

Exercise. Write the matrix

$$\begin{bmatrix} a & b \\ c & d \end{bmatrix} = \alpha_0 \begin{bmatrix} 1 & 0 \\ 0 & 1 \end{bmatrix} + \alpha_1 \begin{bmatrix} 0 & -1 \\ 1 & 0 \end{bmatrix} + \alpha_2 \begin{bmatrix} 0 & -\mathrm{i} \\ -\mathrm{i} & 0 \end{bmatrix} + \alpha_3 \begin{bmatrix} \mathrm{i} & 0 \\ 0 & -\mathrm{i} \end{bmatrix}. \tag{2.5}$$

and evaluate the α_i. Also prove (2.3).

If a matrix Q contains quaternion units, i.e., if $Q = [q_{ij}]$, then the dual of the quaternion matrix is defined by

$$Q^{\mathrm{D}} := [q_{ji}^{\mathrm{D}}] = -ZQ^{\mathrm{t}}Z, \qquad {}^{\mathrm{t}} = \text{transpose}. \tag{2.6}$$

A quaternion matrix is self-dual/anti-self-dual if $Q = \pm Q^{\mathrm{D}} = \mp ZQ^{\mathrm{t}}Z$.

The definition of quaternion matrix makes definition of the quaternion determinant vague, as quaternion elements, in general, are noncommutative. Thus the order of multiplication becomes important. Following the definition in [26, 52, 59], we define the quaternion determinant (denoted by Qdet) of a self-dual matrix Q as

$$\begin{aligned} \operatorname{Qdet} Q &= \sum_P \epsilon_P (q_{j_1} q_{j_2} q_{j_3} \cdots q_{j_r})^{(0)} (q_{k_1k_2} q_{k_2k_3} \cdots q_{k_sk_1})^{(0)} \cdots \\ &= \sum_P (-1)^{n-l} \prod_1^l (q_{ab} q_{bc} \cdots q_{da})^{(0)}, \end{aligned} \tag{2.7}$$

where P is any permutation of indices $(1, 2, \ldots, n)$ consisting of l exclusive cycles of the form

$$(a \to b \to c \to \cdots \to d \to a) \tag{2.8}$$

and $\epsilon_P = (-1)^{n-l}$ is the parity of P. Furthermore, the product of elements in a particular cycle depends on the index with which one begins the cycle, and the value of the whole product on the right-hand side depends on the order in which the l cyclic factors are written. These ambiguities are removed by taking the scalar part of each cyclic factor and we indicate this fact by a superscript zero on each cycle in (2.7).

Now,we state without proof two theorems of Dyson, which are useful in random matrix theory. For proofs, see [52].

Theorem 1. *If we consider an $n \times n$ self-dual matrix Q and the corresponding $2n \times 2n$ matrix $C(Q)$ (obtained by replacing the quaternion units by their 2×2 matrix representations (2.4)) of twice the size, then*

$$\boxed{\det C(Q) = (\operatorname{Qdet} Q)^2.} \tag{2.9}$$

Theorem 2. *Consider n real variables $x_1, x_2, \ldots, x_n$. Let the quaternion elements of the $n \times n$ matrix Q_n satisfy the following properties:*

(1) *The matrix Q_n is self-dual.*

(2) *The quaternion elements of the matrix satisfy*

$$q_{j,k} = f(x_j, x_k) = q_{k,j}^{\mathrm{D}} \tag{2.10}$$

i.e., $q_{j,k}$ depends only on x_j and x_k.

(3) *Moreover, let*

$$\int_{\mathbb{R}} f(x, x)\, \mathrm{d}\mu(x) = c, \tag{2.11}$$

$$\int_{\mathbb{R}} f(x, y) f(y, z)\, \mathrm{d}\mu(x) = f(x, z) + g(x, z) \tag{2.12}$$

where

$$g(x, y) = \lambda f(x, y) - f(x, y)\lambda \tag{2.13}$$

where $\mathrm{d}\mu$ *is a suitable measure,* c *a constant scalar (i.e., multiple of* $\mathbf{1}$*) and* λ *a constant quaternion. Then the theorem states:*

$$\boxed{\int_{\mathbb{R}} \det Q_n \,\mathrm{d}\mu(x_n) = (c-n+1)\det Q_{n-1}} \tag{2.14}$$

where $\det Q_{n-1}$ *is the* $(n-1)\times(n-1)$ *matrix obtained from* Q_n *by removing the row and column containing the variable* x_n*. We can apply this theorem repeatedly to integrate* $\det Q_n$ *over as many variables as we like.*

This theorem is used to calculate the n-point correlation function (1.14) provided we can express the joint-probability distribution of the eigenvalues in terms of determinants of self-dual quaternion matrices, with elements satisfying properties (1) – (3). Dyson has shown that the joint-probability distribution of eigenvalues for the three types of ensembles can be expressed in terms of such quaternion determinants [27].

2. $P_{\beta,N}(x_1,\ldots,x_N)$ as quaternion determinant

In order to use (2.14) in computing multiple integrals (1.14) for the correlation function, we express $P_{\beta,N}(x_1,\ldots,x_N) = c_{\beta,N}|\Delta(x_1,\ldots,x_N)|^\beta \prod w(x_i)$ in terms of determinants ($\beta = 2$) or quaternion determinants ($\beta = 1,4$). We show that conditions (1) – (3) ensure that they constitute kernel functions involving products of orthogonal/skew-orthogonal polynomials.

PROOF FOR $\beta = 2$. Let $p_j(x)$ be "any" polynomial of order j, and $w(x)$ a function which is positive definite and has finite moments in the range $[a,b]$, i.e.,

$$\int_a^b x^k w(x)\,\mathrm{d}x < \infty, \qquad k = 0,1,2,\ldots . \tag{2.15}$$

From the property of Vandermonde's determinant, one can write

$$\det[p_j(x_k)] = \text{const.} \prod_{j<k}(x_k - x_j), \qquad j,k \in \mathbb{N}. \tag{2.16}$$

Also from (2.16) we get

$$\begin{aligned}\prod w(x_j)\prod_{j<k}(x_j-x_k)^2 &= \text{const.}\det[p_j(x_k)]\det\left[p_j(x_k)\prod w(x_j)\right] \\ &= \det[S_N^{(2)}(x_j,x_k)]_{j,k=1,\ldots,N}\end{aligned} \tag{2.17}$$

where

$$S_N^{(2)}(x,y) = w(x)\sum_{j=0}^{N-1}(h_j)^{-1}p_j(x)p_j(y), \tag{2.18}$$

or more symmetrically,

$$S_N^{(2)}(x,y) = \sqrt{w(x)w(y)}\sum_{j=0}^{N-1}(h_j)^{-1}p_j(x)p_j(y). \tag{2.19}$$

Now, we show that if we start with the aim of writing Δ^2 in terms of "any" polynomial, with Dyson – Mehta conditions (2.20), (2.21) imposed on the kernel, the most natural choice of the polynomial is that it is orthonormal (1.1).

We know that $S_N^{(2)}(x,y)$ should satisfy

(1) the normalization condition:

$$\int_a^b S_N^{(2)}(x,x)\,\mathrm{d}x = c^{(2)}, \qquad c^{(2)} \in \mathbb{R},\ c^{(2)} > 0. \tag{2.20}$$

(2) and the convolution property:

$$\int_a^b S_N^{(2)}(x,y) S_N^{(2)}(y,z)\,\mathrm{d}y = S_N^{(2)}(x,z). \tag{2.21}$$

For the Dyson – Mehta theorem to work, $S_N^{(2)}(x,x)$ should correspond to the level density of UE of $N \times N$ random matrices, with $c^{(2)} = N$. Thus conditions (2.20) and (2.21) demand that $p_j(x) \equiv P_j(x)$ are orthogonal polynomials defined in (1.1). □

Proof for $\beta = 4$. Let $t_j(x)$ be "any" polynomial of order j, $t_j'(x)$ its derivative and let $w(x)$ be a function which is positive definite in the range $[a,b]$ and has finite moments (2.15).

Moreover, $P_{4,2N}(x_1,\ldots,x_{2N})$ can be expressed in terms of the quaternion determinant of a kernel function $Q_{2N}^{(4)}$, where $S_{2N}^{(4)}(x_j,x_k)$, $D_{2N}^{(4)}(x_j,x_k)$ and $I_{2N}^{(4)}(x_j,x_k)$ are defined in (1.22):

$$\begin{aligned}(\mathrm{Qdet}[Q_{2N}^{(4)}(x_j,x_k)])^2 &= \det\begin{bmatrix} S_{2N}^{(4)}(x_j,x_k) & D_{2N}^{(4)}(x_j,x_k) \\ I_{2N}^{(4)}(x_j,x_k) & S_{2N}^{(4)\dagger}(x_j,x_k)\end{bmatrix} \\ &= \det\left[\begin{bmatrix}\phi_j(x_k) \\ \psi_j(x_k)\end{bmatrix} Z \begin{bmatrix}\psi_j(x_k) & -\phi_j(x_k)\end{bmatrix}\right] \\ &= \left(\det\begin{bmatrix}\phi_j(x_k) & \psi_j(x_k)\end{bmatrix}\right)^2 \\ &= \left(\text{const.}\prod_{j<k}(x_j-x_k)^4\right)^2.\end{aligned} \tag{2.22}$$

Here $\phi_j(x) = g_j^{-1/2} w(x) t_j(x)$ is the weighted polynomial and $\psi_j(x)$ is its derivative. At this point, g_j is a constant, which is positive definite and real. Finally we show that this is the normalization constant for the skew-orthogonal polynomial.

In the matrices in the second line of the above equation j and k take the values $j = 1,2,\ldots,2N$; $k = 1,2,\ldots,N$. Collecting the results, we have

$$\prod_{j<k}(x_j-x_k)^4 = \text{const.}\,\mathrm{Qdet}\left[Q_{2N}^{(4)}(x_j,x_k)\right], \tag{2.23}$$

where

$$\begin{aligned} Q_{2N}^{(4)}(x,y) = {}&\frac{1}{2}\big(S_{2N}^{(4)}(x,y) + S_{2N}^{(4)}(y,x)\big) + \frac{1}{2}\big(I_{2N}^{(4)}(x,y) - D_{2N}^{(4)}(x,y)\big)e_1 \\ &+ \frac{\mathrm{i}}{2}\big(I_{2N}^{(4)}(x,y) + D_{2N}^{(4)}(x,y)\big)e_2 - \frac{\mathrm{i}}{2}\big(S_{2N}^{(4)}(x,y) - S_{2N}^{(4)}(y,x)\big)e_3, \end{aligned} \tag{2.24}$$

$\mathrm{i} = \sqrt{-1}$ and e_1, e_2, e_3 are quaternion units defined in (2.1) – (2.4). □

Exercise. (1) Show that $Q_{2N}^{(4)}$ may also be conveniently written in its 2×2 matrix representation as

$$Q_{2N}^{(4)}(x,y) = \begin{bmatrix} S_{2N}^{(4)}(x,y) & D_{2N}^{(4)}(x,y) \\ I_{2N}^{(4)}(x,y) & S_{2N}^{(4)\dagger}(x,y) \end{bmatrix}. \tag{2.25}$$

(2) Show that

$$\left[Q_{2N}^{(4)}(x,y)\right]^{\mathrm{D}} = Q_{2N}^{(4)}(y,x).$$

Now, we will show that if we start with the aim of writing Δ^4 in terms kernels with "any" polynomial, the Dyson – Mehta conditions (2.26) and (2.27) restrict them to skew-orthogonal polynomials satisfying (1.2).

From the Dyson – Mehta theorem, $Q_{2N}^{(4)}(x,y)$ should satisfy

(1) the convolution property:

$$\int_a^b Q_{2N}^{(4)}(x,y) Q_{2N}^{(4)}(y,z)\, \mathrm{d}y = Q_{2N}^{(4)}(x,z). \tag{2.26}$$

(2) and the normalization condition:

$$\int_a^b Q_{2N}^{(4)}(x,x)\, \mathrm{d}x = \int \begin{bmatrix} S_{2N}^{(4)}(x,x) & 0 \\ 0 & S_{2N}^{(4)}(x,x) \end{bmatrix} \mathrm{d}x = c^{(4)}, \qquad c^{(4)} \in \mathbb{R},\ c^{(4)} > 0. \tag{2.27}$$

For the Dyson – Mehta theorem to work, $S_{2N}^{(4)}(x,x)$ should correspond to the level density of SE of $2N \times 2N$ random matrices, with $c^{(4)} = N$.

PROOF OF (2.26). Let us define

$$\int_a^b f(x,y) f(y,z)\, \mathrm{d}y := f * f. \tag{2.28}$$

We have

$$\begin{aligned} S_{2N}^{(4)} * S_{2N}^{(4)} + D_{2N}^{(4)} * I_{2N}^{(4)} &= \sum Z_{jk} Z_{lm} \phi_j'(x) \phi_m(y) \big(\phi_k(y) * \phi_l'(x) - \phi_l(x) * \phi_k'(y)\big) \\ &= \sum Z_{jk} Z_{kl} Z_{lm} \phi_j'(x) \phi_m(y) \\ &= \sum (Z^3)_{jm} \phi_j'(x) \phi_m(y) \\ &= S_{2N}^{(4)}. \end{aligned} \tag{2.29}$$

where we have used $Z^2 = -1$.

$$\begin{aligned} S_{2N}^{(4)} * D_{2N}^{(4)} + D_{2N}^{(4)} * S_{2N}^{(4)\dagger} &= -\sum Z_{jk} Z_{lm} \phi_j'(x) \phi_m(y) \big(\phi_k(y) * \phi_l'(x) - \phi_k'(y) * \phi_l(x)\big) \\ &= \sum Z_{jk} Z_{kl} Z_{lm} \phi_j'(x) \phi_m'(y) \\ &= \sum (Z^3)_{jm} \phi_j'(x) \phi_m'(y) \\ &= D_{2N}^{(4)}. \end{aligned} \tag{2.30}$$

$$(2.31)\quad \begin{aligned} I_{2N}^{(4)} * S_{2N}^{(4)} + S_{2N}^{(4)\dagger} * I_{2N}^{(4)} &= \sum Z_{jk} Z_{lm} \phi_j(x)\phi_m(y)\big(\phi_k(y) * \phi_l'(x) - \phi_l(x) * \phi_k'(y)\big) \\ &= \sum Z_{jk} Z_{kl} Z_{lm} \phi_j(x)\phi_m(y) \\ &= \sum \left(Z^3\right)_{jm} \phi_j(x)\phi_m(y) \\ &= I_{2N}^{(4)}. \end{aligned}$$

$$(2.32)\quad \begin{aligned} I_{2N}^{(4)} * D_{2N}^{(4)} + S_{2N}^{(4)\dagger} * S_{2N}^{(4)\dagger} &= \sum Z_{jk} Z_{lm} \phi_j(x)\phi_m'(y)\big(-\phi_k(y) * \phi_l'(x) + \phi_l(x) * \phi_k'(y)\big) \\ &= \sum Z_{jk} Z_{kl} Z_{lm} \phi_j'(x)\phi_m(y) \\ &= -\sum \left(Z^3\right)_{jm} \phi_j(x)\phi_m'(y) \\ &= S_{2N}^{(4)\dagger}. \end{aligned}$$

Using them, it is easy to verify (2.26). □

Conditions (2.26) and (2.27) demand that the skew-orthogonal functions

$$(2.33)\quad \phi_j(x) \equiv \phi_j^{(4)}(x) = g_j^{-1/2} w(x) \pi_j^{(4)}(x), \qquad \psi_j^{(4)}(x) = \frac{\mathrm{d}}{\mathrm{d}x}\phi_j^{(4)}(x),$$

should satisfy

$$(2.34)\quad \begin{aligned} &\int_a^b g_j^{-1}\left[\pi_j^{(4)}(x)\pi_k^{(4)\prime}(x) - \pi_k^{(4)}(x)\pi_j^{(4)\prime}(x)\right] w^2(x)\,\mathrm{d}x \\ &\qquad = \int_a^b \left[\phi_j^{(4)}(x)\psi_k^{(4)}(x) - \phi_k^{(4)}(x)\psi_j^{(4)}(x)\right]\,\mathrm{d}x = Z_{jk}, \end{aligned}$$

with $g_j \equiv g_j^{(4)}$ and Z defined in (1.3).

An alternate definition would be

$$(2.35)\quad \begin{aligned} &\frac{1}{2}\int_a^b {g_j^{(4)}}^{-1}\left[\pi_j^{(4)}(x)\pi_k^{(4)\prime}(x) - \pi_k^{(4)}(x)\pi_j^{(4)\prime}(x)\right] w^2(x)\,\mathrm{d}x \\ &\qquad = \frac{1}{2}\int_a^b \left[\phi_j^{(4)}(x)\psi_k^{(4)}(x) - \phi_k^{(4)}(x)\psi_j^{(4)}(x)\right]\,\mathrm{d}x \\ &\qquad = \int_a^b \phi_j^{(4)}(x)\psi_k^{(4)}(x)\,\mathrm{d}x = Z_{j,k} \end{aligned}$$

where the last step is obtained by integrating by parts, with $\phi_j^{(4)}(a) = \phi_j^{(4)}(b) = 0$.

Here, Kramer's degeneracy is embedded in the definition of the skew-orthogonal functions (2.34). *This is removed in* (2.35) *by integrating by parts, so that the resultant level density normalizes to* $2N$ *instead of* N.

We have used (2.35) in our derivation of the generalized Christoffel – Darboux sum [36] and the differential system [37]. Due to the removal of Kramer's degeneracy, as well as putting the weight functions on an equal footing, this is useful in *proving duality* between skew-orthogonal functions arising in the study of orthogonal and symplectic ensembles of random matrices. However, (2.35) differs from that of [35, 40, 53] for $\beta = 4$ by a factor of 2. Obviously, this will create confusion in our definition of the level density, which has to be divided by a factor of 2. We will encounter this in Chapter 6 while studying the level density of the quartic double well potential.

Note. Skew-orthogonal polynomials $\pi_j^{(4)}(x)$ can also be defined with respect to $w(x)$, instead of $w^2(x)$. This definition is used in [35] and [40] and is also used in this book to study statistical properties of symplectic ensembles.

PROOF FOR $\beta = 1$. Let $q_j(x)$ be "any" polynomial of order j. Let $\phi_j(x) = w(x)q_j(x)$ and $\psi_j^{(1)}(x) = \int \phi_j(y)\epsilon(x-y)\,dy$ and $w(x)$ be function which is positive definite, defined in the range $[a,b]$ and with finite moments (2.15).

Moreover, $P_{1,2N}(x_1,\ldots,x_N)$ can be expressed in terms of quaternion determinants of a kernel function $Q_{2N}^{(1)}$, where $S_{2N}^{(1)}(x_j,x_k)$ $D_{2N}^{(1)}(x_j,x_k)$ and $I_{2N}^{(1)}(x_j,x_k)$ are defined in (1.22).

We note that the $2N\times 2N$ matrix

$$G = \det\begin{bmatrix} S_N^{(1)}(x_j,x_k) & D_N^{(1)}(x_j,x_k) \\ I_N^{(1)}(x_j,x_k) & S_N^{(1)\dagger}(x_j,x_k)\end{bmatrix}_{j,k=1,2,\ldots,N} = \det\left[\begin{bmatrix}\phi_j(x_k)\\ \psi_j(x_k)\end{bmatrix} Z\begin{bmatrix}\psi_j(x_k) & -\phi_j(x_k)\end{bmatrix}\right]_{k=1,\ldots,N;j=1,\ldots,2N} \tag{2.36}$$

has rank N. Thus the N rows $\left[I_N^{(1)}(x_j,x_k) \quad S_N^{(1)\dagger}(x_j,x_k)\right]$ are linear combinations of the N rows $\left[S_N^{(1)}(x_j,x_k) \quad D_N^{(1)}(x_j,x_k)\right]$, and therefore $\det C([Q_N^{(1)}(x_j,x_k)])$ will not be changed if we subtract from its last N rows the last N rows of G,

$$\det C([Q_N^{(1)}(x_j,x_k)]) = \det[\epsilon(x_j-x_k)]\det[D_N^{(1)}(x_j,x_k)]. \tag{2.37}$$

Now

$$\det[D_N^{(1)}(x_j,x_k)] = \det Z\cdot(\det[\phi_j(x_k)])^2 = \text{const.}\prod_{j<k}(x_j-x_k)^2 \tag{2.38}$$

and

$$\det[\epsilon(x_j-x_k)] = 1. \tag{2.39}$$

Collecting the results one gets

$$\det C[Q_N^{(1)}(x_j,x_k)] = \text{const.}\prod_{j<k}(x_j-x_k)^2, \tag{2.40}$$

or equivalently

$$Q\det[Q_N^{(1)}(x_j,x_k)] = \text{const.}\prod_{j<k}|x_j-x_k|, \tag{2.41}$$

where, $C([Q_N^{(1)}])$ is the $2N\times 2N$ version of the quaternion matrix $Q_N^{(1)}$.

Also the kernel $Q_N^{(1)}(x,y)$, with $J_N^{(1)}(x,y) = I_N^{(1)}(x,y) - \epsilon(x-y)$ is given by

$$\begin{aligned} Q_N^{(1)}(x,y) &= \frac{1}{2}(S_N^{(1)}(x,y) + S_N^{(1)\dagger}(x,y)) + \frac{1}{2}(J_N^{(1)}(x,y) - D_N^{(1)}(x,y))e_1 \\ &\quad + \frac{\mathrm{i}}{2}(J_N^{(1)}(x,y) + D_N^{(1)}(x,y))e_2 - \frac{\mathrm{i}}{2}(S_N^{(1)}(x,y) - S_N^{(1)\dagger}(x,y))e_3. \end{aligned} \tag{2.42}$$

where $\mathrm{i} = \sqrt{-1}$ and e_1, e_2, e_3 are quaternion units defined in (2.1)–(2.4).

Exercise. (1) Show that $Q_N^{(1)}$ may also be conveniently written in its 2×2 matrix representation as

$$Q_N^{(1)}(x,y) = \begin{bmatrix} S_N^{(1)}(x,y) & D_N^{(1)}(x,y) \\ J_N^{(1)}(x,y) & S_N^{(1)\dagger}(x,y) \end{bmatrix}. \tag{2.43}$$

(2) From the definition of dual of a matrix (2.6), show

$$[Q_N^{(1)}(x,y)]^{\mathrm{D}} = Q_N^{(1)}(y,x).$$

Now, we show that if we start with the aim of writing $|\Delta|$ in terms of kernels with "any" polynomial, the Dyson – Mehta conditions (2.44), (2.45) restricts them to skew-orthonormal polynomials satisfying (1.2).

From the Dyson – Mehta theorem, $Q_{2N}^{(1)}(x,y)$ should satisfy

(1) the normalization condition

$$\int_a^b Q_N^{(1)}(x,x)\,\mathrm{d}x = \int_a^b \begin{bmatrix} S_N^{(1)}(x,x) & 0 \\ 0 & S_N^{(1)}(x,x) \end{bmatrix} \mathrm{d}x = c^{(1)}, \tag{2.44}$$

$$c^{(1)} \in \mathbb{R},\ c^{(1)} > 0,$$

(2) and the convolution condition

$$\int_a^b Q_N^{(1)}(x,y) Q_N^{(1)}(y,z)\,\mathrm{d}y = Q_N(x,z) + g(x,z) \tag{2.45}$$

where

$$g(x,z) = Q_N^{(1)}(x,z)\lambda - \lambda Q_N^{(1)}(x,z)$$

and

$$\lambda = \frac{\mathrm{i}}{2} e_3.$$

For the Dyson – Mehta theorem to work, $S_N^{(1)}(x,x)$ correspond to the level density of OE of $N \times N$ random matrices, with $c^{(1)} = N$.

Following a method similar to that used in the SE, we find that conditions (2.44) and (2.45) demand that $\phi_j(x) \equiv \phi_j^{(1)}(x) = \left(g_j^{(1)}\right)^{-1/2} w(x) \pi_j^{(1)}(x)$, should satisfy the skew-orthonormality property

$$\int_a^b \phi_j^{(1)}(x) \psi_k^{(1)}(x)\,\mathrm{d}x = Z_{jk}, \qquad j,k \in \mathbb{N}, \tag{2.46}$$

where $Z = \mathrm{i}\sigma_2$ is the antisymmetric, anti-unitary matrix defined in (1.3). Also $w(x)$ is the weight function defined in the interval $[a,b]$ and $g_j^{(1)}$ is the normalization constant. □

3. Gaussian ensemble results for correlation functions

Bulk of the spectrum. We summarize here the results for $R_n^{(\beta)}$, first obtained for the Gaussian ensembles [26, 27, 51, 53]. We have for the three types of ensembles

$$R_n^{(\beta)}(r_1, \ldots, r_n) = \mathrm{Qdet}[\sigma_\beta(r_j - r_k)]_{j,k=1,\ldots,n} = \{\det[\sigma_\beta(r_j - r_k)]\}^{1/2}, \tag{2.47}$$

where

$$\sigma_2(r) = \begin{pmatrix} S(r) & 0 \\ 0 & S(r) \end{pmatrix}, \tag{2.48}$$

$$\sigma_1(r) = \begin{pmatrix} S(r) & D(r) \\ I(r) - \epsilon(r) & S(r) \end{pmatrix}, \tag{2.49}$$

$$\sigma_4(r) = \begin{pmatrix} S(2r) & D(2r) \\ I(2r) & S(2r) \end{pmatrix}, \tag{2.50}$$

$$S(r) = \sin(\pi r)/\pi r, \tag{2.51}$$

$$D(r) = \mathrm{d}S(r)/\,\mathrm{d}r, \tag{2.52}$$

$$I(r) = \int_0^r S(r')\,\mathrm{d}r', \tag{2.53}$$

$$\epsilon(r) = r/2|r|. \tag{2.54}$$

The function $S(r)$, from which all other functions above derive, will be seen in the following chapters as the asymptotic form of a kernel function $S_N^{(\beta)}(x,y)$ involving orthogonal ($\beta = 2$) and skew-orthogonal polynomials ($\beta = 1, 4$). Universality of $R_n^{(\beta)}$ will follow from that of the asymptotic kernel function $S(r)$. As an example of (2.47), we give below the $R_2^{(\beta)}$ for $\beta = 1$, 2 and 4:

$$R_2^{(1)} = 1 - [S(r)]^2 + J(r)D(r), \tag{2.55}$$

$$R_2^{(2)} = 1 - [S(r)]^2, \tag{2.56}$$

$$R_2^{(4)} = 1 - [S(2r)]^2 + I(2r)D(2r), \tag{2.57}$$

where $J(r) = I(r) - \epsilon(r)$.

4. Level density: An alternate approach

In this section we use a nonpolynomial method to derive level densities of random matrix ensembles as $N \to \infty$.

From the definition of $P_{\beta,N}(x_1,\ldots,x_N)$, we have

$$\frac{\partial P_{\beta,N}(x_1,\ldots,x_N)}{\partial x_1} = \left(\beta \sum_{j\neq 1} \frac{1}{x_1 - x_j} + \frac{w'(x_1)}{w(x_1)}\right) P_{\beta,N}(x_1,\ldots,x_N). \tag{2.58}$$

This along with equation (1.14) gives us an exact hierarchical set of relations linking $R_n^{(\beta)}(x_1,\ldots,x_n)$ to $R_{n+1}^{(\beta)}(x_1,\ldots,x_{n+1})$ [33, 66, 68]. We suppress β to avoid cluttering. For $n = 1$, this gives

$$\frac{\partial R_1(x)}{\partial x} = \beta \int \frac{R_2(x,y)}{x-y}\,\mathrm{d}y + \frac{w'(x)}{w(x)} R_1(x). \tag{2.59}$$

For large N, the integral on the right side can be replaced by a principal-value integral involving $R_2(x,y) \approx R_1(x)R_1(y)$; moreover $\partial R_1(x)/\partial x$ can be dropped. Both these approximations can be rigorously justified from the behavior of $R_1(x)$ and $R_2(x,y)$ for large N. We thus find [4, 27, 66]

$$\beta R_1(x) \int \frac{R_1(y)}{x-y}\,\mathrm{d}y + \frac{w'(x)}{w(x)} R_1(x) = 0, \qquad w(x) \neq 0, \tag{2.60}$$

a result which could also be derived directly by maximizing $\log P_{\beta,N}$. We note that for weight functions of the Jacobi and Laguerre type, $w'(x)/w(x)$ gives rise to singularities at the end points which have to be removed, while for a polynomial potential they appear as moments of the density function which can be interpreted as an ordered parameter defining the birth of a cut.

We solve below the integral equation (2.60), using the resolvent

$$G(z) = \int \frac{R_1(y)}{z-y}\,\mathrm{d}y, \tag{2.61}$$

which satisfies

$$G(x+\mathrm{i}0) = \int \frac{R_1(y)}{x-y}\,\mathrm{d}y - \mathrm{i}\pi R_1(x). \tag{2.62}$$

Jacobi weight. For Jacobi weight function (1.6), defined in the range $[-1,1]$, one needs to consider carefully the singularities of $w'(x)/w(x)$ at $x=\pm 1$. Since $R_1(x)=0$ for $|x|>1$, and $O(N)$ for $|x|\le 1$, we find from (2.60), after multiplication by $(1-x^2)/(z-x)$ and integration over x, that

$$\int_{-1}^{1} \mathrm{d}x\,\frac{(1-x^2)R_1(x)}{z-x}\int_{-1}^{1}\mathrm{d}y\,\frac{R_1(y)}{x-y} = 0,$$

the terms ignored being of lower order in N. The expression on the left can be written as

$$\begin{aligned} &\frac{1}{2}\int\int_{-1}^{1}\mathrm{d}x\,\mathrm{d}y\,\frac{R_1(x)R_1(y)}{x-y}\left(\frac{1-x^2}{z-x}-\frac{1-y^2}{z-y}\right) \\ &= \frac{1}{2}\int\int_{-1}^{1}\mathrm{d}x\,\mathrm{d}y\,\frac{R_1(x)R_1(y)}{(z-x)(z-y)}\{1-z(x+y)+xy\} \\ &= \frac{1}{2}\int\int_{-1}^{1}\mathrm{d}x\,\mathrm{d}y\,\frac{R_1(x)R_1(y)}{(z-x)(z-y)}\{1-z^2+(z-x)(z-y)\} \\ &= \frac{1}{2}(1-z^2)G^2(z)+\frac{N^2}{2}. \end{aligned} \tag{2.63}$$

Thus, since $G(z)\approx N/z$ for large $|z|$, we get

$$G(z) = \frac{N}{\sqrt{z^2-1}}, \tag{2.64}$$

and

$$R_1(x) = \begin{cases} \dfrac{N}{\pi\sqrt{1-x^2}}, & |x|<1, \\ 0, & |x|>1, \end{cases} \tag{2.65}$$

the result being the same for all finite values of the parameters a, b. Note that the level density becomes indefinitely large at the end points.

Associated Laguerre weight. For weight functions defined in the range $[0,\infty]$, and of the form

$$w(x) = x^{\alpha}\exp[-\beta V(x)], \qquad \alpha>-1,\ V(x)=\sum_{k=1}^{2d-1}\frac{u_k x^k}{k}; \tag{2.66}$$

$R_1(x)=0$ for $x<0$, and $w'(x)/w(x)$ has a singularity at $x=0$. In this case we find from (2.60), after multiplication by $xV'(x)/(z-x)$, and integration over x, that

$$-\int_0^{\infty}\frac{xV'(x)R_1(x)}{z-x}\,\mathrm{d}x + \int_0^{\infty}\mathrm{d}x\,\frac{xR_1(x)}{z-x}\int_0^{\infty}\mathrm{d}y\frac{R_1(y)}{x-y} = 0, \tag{2.67}$$

so that

$$zG^2(z) - 2zV'(z)G(z) + \int \frac{zV'(z) - xV'(x)}{z-x} R_1(x)\ dx = 0, \tag{2.68}$$

giving

$$\pi^2 x[R_1(x)]^2 = 2\int \left[\frac{xV'(x) - yV'(y)}{x-y}\right] R_1(y)\ dy - x[V'(x)]^2. \tag{2.69}$$

Exercise. Show that for associated Laguerre weight (1.7), the level density is given by

$$R_1^{(\beta)}(x) = \frac{1}{\pi}\sqrt{\frac{2N-x}{x}}, \qquad 2N > x > 0. \tag{2.70}$$

This is verified in Chapters 3–6 using the polynomial method.

Polynomial potential. For weight function defined in the range $[-\infty, \infty]$ of the form

$$w(x) = \exp[-\beta N V(x)], \qquad V(x) = \sum_{k=0}^{2d} \frac{u_k x^k}{k}, \tag{2.71}$$

we have from (2.60)

$$\begin{aligned} 0 &= -N\int_{-\infty}^{\infty} \frac{V'(x)R_1(x)}{z-x}\ dx + \int_{-\infty}^{\infty} dx\ \frac{R_1(x)}{z-x}\int_{-\infty}^{\infty} dy\ \frac{R_1(y)}{x-y} \\ &= \frac{1}{2}\int_{-\infty}^{\infty}\int_{-\infty}^{\infty} \frac{R_1(x)R_1(y)}{(x-y)}\left[\frac{1}{(z-x)} - \frac{1}{(z-y)}\right] dx\ dy - NV'(z)G(z) \\ &\qquad + N\int_{-\infty}^{\infty} \frac{(V'(z) - V'(x))}{(z-x)} R_1(x)\ dx \end{aligned} \tag{2.72}$$

implying

$$G^2(z) - 2NV'(z)G(z) + 2N\int_{-\infty}^{\infty} \frac{(V'(z) - V'(x))}{(z-x)} R_1(x)\ dx = 0. \tag{2.73}$$

we have

$$\pi^2 R_1^2(x) = 2N\int_{-\infty}^{\infty} \frac{(V'(z) - V'(x))}{(z-x)} R_1(x)\ dx - N[V'(x)]^2, \tag{2.74}$$

For quartic potential,

$$V(x) = \left(\frac{gx^4}{4} + t\frac{x^2}{2}\right), \qquad g > 0,\ t \in \mathbb{R}, \tag{2.75}$$

there exist two minima for $t \le t_c$ [10]. In this case, (2.74) yields

$$\left(\frac{\pi}{N}\right)^2 [R_1(x)]^2 = 2(gM_2/N + t) + x^2[2g - (gx^2 + t)^2], \tag{2.76}$$

where the kth moment $M_k = \int x^k R_1(x)\ dx$. In the above calculation, we have scaled out β. Hence we can use the method of orthogonal polynomials to calculate M_2 for all β. However, in so doing, we must remember that we get different level densities for different values of β since the weight function changes with β. For example, to recover results of [10], we need to rescale $V(x) \to V(x)/2$. We get the level density

$$\left(\frac{\pi}{N}\right)^2 [R_1(x)]^2 = (gM_2/N + t) + x^2\left[g - \left(\frac{gx^2+t}{2}\right)^2\right]. \tag{2.77}$$

Now, if $M_2 = -tN/g$, then $R_1(0) = 0$, suggesting the possibility of band splitting at $x = 0$; in fact $R_1(x)$ is then zero for $x^2 \le -t/g - (2/\sqrt{g})$ implying a two-band structure for $t \le t_c \equiv -2\sqrt{g}$. To prove this, we evaluate M_2 by the polynomial method. Since M_2 does not depend on β, we consider $\beta = 2$ (along with the scaling mentioned before) for which [27]

$$R_1(x) = \sum_{\mu=0}^{N-1} (h_\mu)^{-1}[P_\mu(x)]^2 \exp[-NV(x)], \tag{2.78}$$

$P_\mu(x)$ being the orthogonal polynomials with normalization h_μ,

$$\int_{-\infty}^{\infty} P_\mu(x)P_\nu(x) \exp[-NV(x)]\ \mathrm{d}x = h_\mu\delta_{\mu\nu}. \tag{2.79}$$

For monic polynomials (i.e., polynomials with highest coefficient unity), we have the three-term recursion relation [77], $xP_\mu = P_{\mu+1} + R_\mu P_{\mu-1}$ where $R_{\mu+1} = h_{\mu+1}/h_\mu$, for $\mu \ge 0$ and $R_0 = 0$. Also we have

$$\begin{aligned} M_2 &:= \int x^2 R_1(x)\ \mathrm{d}x \\ &= \sum_{\mu=0}^{N-1} (h_\mu)^{-1} \int [P_{\mu+1}(x) + R_\mu P_{\mu-1}(x)] \\ &\qquad\qquad \times [P_{\mu+1}(x) + R_\mu P_{\mu-1}(x)] w(x)\ \mathrm{d}x \\ &= \sum_{\mu=0}^{N-1} [R_{\mu+1} + R_\mu]. \end{aligned} \tag{2.80}$$

Also, since $\int \mathrm{d}x\, [P_{\mu+1}(x)P_\mu(x) \exp\{-NV(x)\}]' = 0$, we have

$$NR_{\mu+1}[g(R_{\mu+2} + R_{\mu+1} + R_\mu) + t] = \mu + 1, \tag{2.81}$$

valid for $\mu \ge 0$. It can be shown that there can be two cases:

(i) for large μ, R_μ oscillates, i.e.,

$$R_{\mu+1} \simeq R_{\mu+3} = a_1; \qquad R_\mu \simeq R_{\mu+2} = a_{-1} \tag{2.82}$$

or

(ii) R_μ remains constant, i.e.,

$$R_\mu \simeq R_{\mu\pm1} = a_0. \tag{2.83}$$

Solving (2.81) iteratively for case (i), we see that R_μ alternates between

$$a_{\pm1} = \frac{1}{2g}[-t \pm (t^2 - 4\lambda g)^{1/2}], \quad \lambda = \frac{\mu}{N}, \qquad \text{for } \lambda \le \lambda_c \equiv t^2/2g, \tag{2.84}$$

while R_μ becomes

$$a_0 = \frac{1}{6g}[-t - (t^2 + 12\lambda g)^{1/2}], \qquad \text{for } \lambda \ge \lambda_c \equiv t^2/2g, \tag{2.85}$$

all three values being the same for $\lambda = \lambda_c$. Thus M_2 is given, for $N \to \infty$, by

$$M_2 = \begin{cases} -tN/g, & t \le t_c, \\ -\dfrac{N}{54g^2}[t^3 - (t^2+12g)^{3/2} + 18gt], & t \ge t_c, \end{cases}$$

where the last result is obtained by writing $\sum_0^{N-1} = \sum_0^{\lambda_c N} + \sum_{\lambda_c N}^{N-1}$ for $t < 0$ and using the value of a_0 for $1 \ge \lambda \ge \lambda_c$ in (2.80) and finally replacing summation by an integral. Finally, (2.77) yields, for $t \le t_c$, the two-band density

$$N^{-1}R_1(x) = \frac{|x|}{\pi}\left[g - \left(\frac{gx^2+t}{2}\right)^2\right]^{1/2}, \tag{2.86}$$

and, for $t \ge t_c$, the one-band density

$$N^{-1}R_1(x) = \frac{g}{2\pi}\left[\frac{1}{3g}\{(t^2+12g)^{1/2}+2t\}+x^2\right][(2/3g)\{(t^2+12g)^{1/2}-t\}-x^2]^{1/2}.$$

In both cases, the density is zero when the right-hand side is imaginary. Note that $R_1(x)$ (for the one-band case), which is peaked at $x = 0$ for $t \ge t_c$, develops a minimum at $x = 0$ for $t < t_c$.

Exercise. For the Gaussian weight (1.8), show

$$R_1(x) = \begin{cases} \dfrac{1}{\pi}\sqrt{2N - x^2}, & |x| < \sqrt{2N}, \\ 0, & |x| > \sqrt{2N}. \end{cases} \tag{2.87}$$

The polynomial methods discussed in Chapters 3 – 6 confirm the above results.

Note. The above approach is applicable to a wide variety of polynomial potentials (see for example [41]). The only difficulty lies in the laborious calculations of higher order moments, for which one can look for an appropriate recursion relation.

CHAPTER 3

The $S_N^{(\beta)}(x, y)$ Kernel and Christoffel – Darboux Formulas

In this chapter, we evaluate the sum $S_N^{(\beta)}(x, y)$ for $\beta = 1, 2$ and 4. $S_N^{(2)}(x, y)$ is a sum over products of orthogonal polynomials (1.18) and its solution has been known for decades. It is called the Christoffel – Darboux formula (see for example [77]). We compute $S_N^{(\beta)}(x, y)$, $\beta = 1$ and 4 [36,38,39], which are sums over products of skew-orthogonal functions (1.22). We call them the generalized Christoffel – Darboux formula. Apart from improving our understanding of the theory of skew-orthogonal functions, these results make calculations of OE and SE more transparent compared to the method used so far using Sener – Verbaarschot – Widom formalism [72,78].

1. Method of orthogonal polynomials

The correlation function of the unitary ensemble of random matrices ($\beta = 2$) can be written as (see Chapter 2 for details)

$$R_n^{(2)}(x_1, \ldots, x_n) = \det[S_N^{(2)}(x_j, x_k)]_{j,k=1,\ldots,n}, \tag{3.1}$$

where $S_N^{(2)}(x, y)$ is the kernel function

$$S_N^{(2)}(x, y) = \sum_{j=0}^{N-1} \phi_j^{(2)}(x)\phi_j^{(2)}(y), \qquad \phi_j^{(2)}(x) = \sqrt{\frac{w(x)}{h_j}} P_j(x), \tag{3.2}$$

the $P_j(x)$ are orthogonal polynomials (1.1) of order j, $w(x)$ the weight function with respect to which they are defined, and h_j the normalization constant. The Christoffel – Darboux formula expresses (3.2) as a linear combination of products of only $\phi_N^{(2)}(x)$ and $\phi_{N-1}^{(2)}(x)$. This result is useful in studying the statistical properties of unitary ensembles of random matrices.

2. Christoffel – Darboux formula: $\beta = 2$

We begin this section with an alternative derivation of the Christoffel – Darboux formula (see the classical book of Szegő [77] for the conventional proof).

PROOF. We define semi-infinite matrices $\Phi^{(2)}(x)$, comprising weighted orthogonal polynomials $\phi_n^{(2)}(x)$:

$$\Phi^{(2)}(x) := (\phi_0^{(2)}(x) \ldots \phi_n^{(2)}(x) \ldots)^{\mathrm{t}}, \qquad {}^{\mathrm{t}} = \text{transpose}. \tag{3.3}$$

We write $S_N^{(2)}(x, y)$ of (3.2) as

$$S_N^{(2)}(x, y) = \Phi^{(2)\mathrm{t}}(x) \prod_N \Phi^{(2)}(y), \tag{3.4}$$

where we use the diagonal matrix $\Pi_N = \mathrm{diag}(\underbrace{1,\ldots,1}_{N},0,\ldots,0)$ to make the sum finite.

Orthogonality ensures that they satisfy a three-term recursion relation. This is the same as saying

$$\boxed{x\Phi^{(2)}(x) = Q^{(2)}\Phi^{(2)}(x)} \tag{3.5}$$

where $Q^{(2)}$ is the well-known Jacobi matrix. Using the orthonormality condition, we get

$$Q^{(2)}_{j,k} = \int_{\mathbb{R}} x\phi_j^{(2)}(x)\phi_k^{(2)}(x)\,\mathrm{d}x = Q^{(2)}_{k,j} \implies Q^{(2)} = Q^{(2)\mathrm{t}}. \tag{3.6}$$

From equations (3.4), (3.5), and (3.6), we get

$$\begin{aligned}(x-y)S_N^{(2)}(x,y) &= (x-y)\Phi^{(2)\mathrm{t}}(x)\prod_N\Phi^{(2)}(y)\\ &= \Phi^{(2)\mathrm{t}}(x)Q^{(2)\mathrm{t}}\prod_N\Phi^{(2)}(y) - \Phi^{(2)\mathrm{t}}(x)\prod_N Q^{(2)}\Phi^{(2)}(y)\\ &= \Phi^{(2)\mathrm{t}}(x)\Big[Q^{(2)}\prod_N - \prod_N Q^{(2)}\Big]\Phi^{(2)}(y)\\ &= \Phi^{(2)\mathrm{t}}(x)\Big[Q^{(2)},\prod_N\Big]\Phi^{(2)}(y).\end{aligned} \tag{3.7}$$

Thus we have

$$\boxed{S_N^{(2)}(x,y) = \frac{\Phi^{(2)\mathrm{t}}(x)[Q^{(2)},\Pi_N]\Phi^{(2)}(y)}{(x-y)}.}$$

In terms of the matrix elements, this boils down to

$$\begin{aligned} S_N^{(2)}(x,y) &= Q^{(2)}_{N-1,N}\left(\frac{\phi_N^{(2)}(x)\phi_{N-1}^{(2)}(y) - \phi_N^{(2)}(y)\phi_{N-1}^{(2)}(x)}{x-y}\right),\\ Q^{(2)}_{N-1,N} &= \frac{k_{N-1}}{k_N}\sqrt{\frac{h_N}{h_{N-1}}},\end{aligned} \tag{3.8}$$

where k_N is the coefficient of x^N in $P_N(x)$ and h_N is the normalization constant. □

3. Method of skew-orthogonal polynomials

For an ensemble of $2N \times 2N$ random matrices invariant under the orthogonal ($\beta = 1$), unitary ($\beta = 2$) and symplectic ($\beta = 4$) transformations, we rewrite (1.14) as

$$R_n^{(\beta)}(x_1,\ldots,x_n) = \frac{2N!}{(2N-n)!}\int \mathrm{d}x_{n+1}\cdots\int \mathrm{d}x_{2N}\; P_{\beta,N}(x_1,x_2\ldots,x_{2N}),\qquad n = 1,2,\ldots. \tag{3.9}$$

As discussed in Chapter 2, for $\beta = 1$ and 4, the evaluation of such integrals requires the joint probability distribution $P_{\beta,N}(x_1,x_2,\ldots,x_{2N})$ to be written in terms of quaternion determinants (i.e., determinant of a matrix, each of whose element is a 2×2 quaternion) satisfying certain properties. From this, one can calculate (3.9). For example, the n-point correlation function $R_n^{(\beta)}(x_1,\ldots,x_n)$, and the level density $R_1^{(\beta)}(x)$, ($n = 1$) is given by

$$R_n^{(\beta)}(x_1,\ldots,x_n) = \mathrm{Qdet}[Q_{2N}^{(\beta)}(x_j,x_k)];\qquad j,k = 1,\ldots,n,\; \beta = 1,4, \tag{3.10}$$

where

$$Q_{2N}^{(\beta)}(x,y) = \begin{pmatrix} S_{2N}^{(\beta)}(x,y) & D_{2N}^{(\beta)}(x,y) \\ I_{2N}^{(\beta)}(x,y) - \delta_{1,\beta}\epsilon(x-y) & S_{2N}^{(\beta)}(y,x) \end{pmatrix}; \tag{3.11}$$
$$R_1^{(\beta)}(x) := S_{2N}^{(\beta)}(x,x).$$

Here δ is the Kronecker delta, introduced to write the results of the two invariant classes of ensembles under one banner. In terms of skew-orthogonal functions $\phi_n^{(\beta)}(x)$ and $\psi_n^{(\beta)}(x)$, we rewrite the expressions of (1.22) as

$$S_{2N}^{(\beta)}(x,y) := \sum_{j,k=0}^{2N-1} Z_{j,k}\phi_j^{(\beta)}(x)\psi_k^{(\beta)}(y) = \widehat{\Psi}^{(\beta)}(y)\prod_{2N}\Phi^{(\beta)}(x) = -\widehat{\Phi}^{(\beta)}(x)\prod_{2N}\Psi^{(\beta)}(y), \tag{3.12}$$

$$D_{2N}^{(\beta)}(x,y) := -\sum_{j,k=0}^{2N-1} Z_{j,k}\phi_j^{(\beta)}(x)\phi_k^{(\beta)}(y) = \widehat{\Phi}^{(\beta)}(x)\prod_{2N}\Phi^{(\beta)}(y), \tag{3.13}$$

$$I_{2N}^{(\beta)}(x,y) := \sum_{j,k=0}^{2N-1} Z_{j,k}\psi_j^{(\beta)}(x)\psi_k^{(\beta)}(y) = -\widehat{\Psi}^{(\beta)}(x)\prod_{2N}\Psi^{(\beta)}(y), \tag{3.14}$$

$$S_{2N}^{(\beta)}(y,x) = S_{2N}^{\dagger(\beta)}(x,y) = \widehat{\Phi}^{(\beta)}(x)\prod_{2N}\Psi^{(\beta)}(y). \tag{3.15}$$

where $\phi_n^{(\beta)}(x)$ and $\psi_n^{(\beta)}(x)$ are defined in (1.23) and (1.24) respectively. Here, $\Pi_{2N} = \mathrm{diag}(\underbrace{1,\dots,1}_{2N},0,\dots,0)$ is a diagonal matrix and $\Phi^{(\beta)}(x)$ and $\Psi^{(\beta)}(x)$ are semi-infinite vectors:

$$\Phi^{(\beta)}(x) = (\Phi_0^{(\beta)\mathrm{t}}(x)\dots\Phi_n^{(\beta)\mathrm{t}}(x)\dots)^{\mathrm{t}}, \qquad \widehat{\Phi}^{(\beta)}(x) = -\Phi^{(\beta)\mathrm{t}}(x)Z, \tag{3.16}$$

$$\Psi^{(\beta)}(x) = (\Psi_0^{(\beta)\mathrm{t}}(x)\dots\Psi_n^{(\beta)\mathrm{t}}(x)\dots)^{\mathrm{t}}, \qquad \widehat{\Psi}^{(\beta)}(x) = -\Psi^{(\beta)\mathrm{t}}(x)Z, \tag{3.17}$$

with each entry a 2×1 matrix:

$$\Phi_n^{(\beta)}(x) = \begin{pmatrix} \phi_{2n}^{(\beta)}(x) \\ \phi_{2n+1}^{(\beta)}(x) \end{pmatrix}, \qquad \widehat{\Phi}_n^{(\beta)}(x) = \begin{pmatrix} \phi_{2n+1}^{(\beta)}(x) \\ -\phi_{2n}^{(\beta)}(x) \end{pmatrix}^{\mathrm{t}}. \tag{3.18}$$

(similarly for $\Psi_n^{(\beta)}(x)$ and $\widehat{\Psi}_n^{(\beta)}(x)$). The antisymmetric block-diagonal matrix Z is given in (1.3), such that $Z = -Z^{\mathrm{t}}$ and $Z^2 = -1$.

Following (1.24), we write

$$\Psi_n^{(4)}(x) = \Phi_n^{\prime(4)}(x), \quad \Psi_n^{(1)}(x) = \int_{\mathbb{R}} \Phi_n^{(1)}(y)\epsilon(x-y)\,\mathrm{d}y, \qquad n\in\mathbb{N}. \tag{3.19}$$

These polynomials satisfy skew-orthonormal relations with respect to the weight function $w^2(x)$[1]:

$$(\Phi_n^{(\beta)},\widehat{\Psi}_m^{(\beta)}) \equiv \int_{\mathbb{R}} \Phi_n^{(\beta)}(x)\widehat{\Psi}_m^{(\beta)}(x)\,\mathrm{d}x = \delta_{nm}\begin{pmatrix} 1 & 0 \\ 0 & 1 \end{pmatrix}, \qquad n,m\in\mathbb{N}. \tag{3.20}$$

[1]To observe duality between the two families of polynomials $\pi_n^{(\beta)}(x)$, $\beta = 1,4$, we skew-orthonormalize them with respect to $w^2(x)$ (as in (3.20)) to *set* both the families of skew-orthogonal functions on an equal footing. However, to study the statistical properties of symplectic ensembles only, this is not needed.

Thus we see that skew-orthonormalization (1.2) in ordinary space is the same as orthonormalization in the 2×2 quaternion space (3.20).

Finally, from (3.19) and (3.12), (3.13), and (3.14) we get

$$D_{2N}^{(1)}(x,y) = -\frac{\partial S_{2N}^{(1)}(x,y)}{\partial y}, \qquad S_{2N}^{(1)}(x,y) = \frac{\partial I_{2N}^{(1)}(x,y)}{\partial y}, \tag{3.21}$$

$$I_{2N}^{(4)}(x,y) = \frac{\partial S_{2N}^{(4)}(x,y)}{\partial x}, \qquad S_{2N}^{(4)}(x,y) = -\frac{\partial D_{2N}^{(4)}(x,y)}{\partial x}. \tag{3.22}$$

Thus a knowledge of the kernel function $S_{2N}^{(\beta)}(x,y)$ is enough to calculate the correlation function. In this chapter, we derive the sum $S_{2N}^{(\beta)}(x,y)$ for $\beta = 1, 4$. The large N behavior of $S_{2N}^{(\beta)}(x,y)$ is studied in Chapters 6–8 after we discover a method to compute large N asymptotics of skew-orthogonal functions and their recursion coefficients (Chapter 4).

4. The generalized Christoffel–Darboux formulas: $\beta = 1$ and 4

4.1. Recursion relations for skew-orthogonal functions with general weight. Let us consider skew-orthogonal polynomials defined with respect to the following weight functions:

- polynomials defined in the finite range $[x_1, x_{2d}]$, which we call the generalized Jacobi weight

$$w_{\alpha_i}(x) = (x_1 - x)^{\alpha_1}(x - x_2)^{\alpha_2} \cdots (x_{2d-1} - x)^{\alpha_{2d-1}}(x - x_{2d})^{\alpha_{2d}}, \qquad x_j \in \mathbb{R} \tag{3.23}$$

  where α_i's are suitably adjusted to ensure that they have finite moments (2.15).
- Polynomials defined in the range $[0, \infty]$, which we call the generalized Laguerre weight

$$w_\alpha(x) = x^\alpha \exp[-V(x)], \qquad V(x) = \sum_{k=1}^{2d-1} \frac{u_k x^k}{k}. \tag{3.24}$$

Finally

- polynomials defined in the range $[-\infty, \infty]$, which we call the generalized Gaussian weight

$$w(x) = \exp[-V(x)], \qquad V(x) = \sum_{k=1}^{2d} \frac{u_k x^k}{k}. \tag{3.25}$$

Evaluation of $\psi_n^{(4)}(x)$ and $\phi_n^{(1)}(x)$ involves terms like $w'_{\alpha_i}(x)/w_{\alpha_i}(x)$ which for (i) have singularity at $x_1, x_2, \ldots, x_{2d}$ and for (ii) have singularity at $x = 0$. Hence we define

$$f(x) := (x - x_1)(x_2 - x)\cdots(x_{2d-1} - x)(x - x_{2d}), \quad \text{for generalized Jacobi} \tag{3.26}$$

$$:= x, \quad \text{for generalized Laguerre,} \tag{3.27}$$

$$:= 1, \quad \text{for generalized Gaussian.} \tag{3.28}$$

To obtain recursion relations, we expand $[f(x)\phi_j^{(\beta)}(x)]'$ and $[xf(x)\phi_j^{(\beta)}(x)]'$ in terms of skew-orthogonal functions $\phi_j^{(\beta)}(x)$. We have

$$\begin{aligned} \left(f(x)\phi_j^{(\beta)}(x)\right)' &= \sum_{k=0}^{j+2d-1} A_{j,k}\phi_k^{(\beta)}(x), \\ \left(xf(x)\phi_j^{(\beta)}(x)\right)' &= \sum_{k=0}^{j+2d} B_{j,k}\phi_k^{(\beta)}(x). \end{aligned} \tag{3.29}$$

It is at this point that we introduce the semi-infinite matrices $P^{(\beta)}$ and $R^{(\beta)}$. The semi-infinite vectors $\Phi^{(\beta)}(x)$ and $\Psi^{(\beta)}(x)$ are given in (3.16) and (3.17) respectively. Using them, we have

$$\boxed{f(x)\Psi^{(4)}(x) = P^{(4)}\Phi^{(4)}(x), \qquad xf(x)\Psi^{(4)}(x) = R^{(4)}\Phi^{(4)}(x),} \tag{3.30}$$

$$\boxed{f(x)\Phi^{(1)}(x) = P^{(1)}\Psi^{(1)}(x), \qquad xf(x)\Phi^{(1)}(x) = R^{(1)}\Psi^{(1)}(x).} \tag{3.31}$$

Equation (3.31) is obtained by multiplying the above expansion by $\epsilon(y-x)$ and integrating by parts. We show that the semi-infinite matrices $P^{(\beta)}$ and $R^{(\beta)}$ have d quaternion bands above and below the central diagonal (hence a matrix with $2d+1$ quaternion bands).

For random matrix analysis, we consider the rather simple cases of Jacobi (1.6) weight function for (i), associated Laguerre (1.7) for (ii) and Gaussian (1.8) and the quartic weight (1.9) for (iii). We show that in the 2×2 quaternion space, the corresponding skew-orthogonal functions satisfy three-term recursion relations for the classical weight functions and $(2d+1)$-term recursion relations for the generalized weight functions.

Recursion relations for skew-orthogonal functions with classical weights. For the classical weights ($d = 1$), $P^{(\beta)}$ and $R^{(\beta)}$ are semi-infinite tridiagonal quaternion matrices. Now, $w'(x)/w(x)$ and hence $\Psi^{(4)}(x)$ and $\Phi^{(1)}(x)$ has singularity at $x \pm 1$ for Jacobi, and at $x = 0$ for associated Laguerre. To remove them, we have

$$f(x) = (1-x^2), \qquad \text{Jacobi} \tag{3.32}$$

$$= x, \qquad \text{Associated Laguerre} \tag{3.33}$$

$$= 1, \qquad \text{Gaussian.} \tag{3.34}$$

In other words, for classical weights, skew-orthogonal functions satisfy three-term recursion relations in quaternion space and is given by

$$f(x)\Psi_n^{(4)}(x) = \mathbf{P}_{n,n+1}^{(4)}\Phi_{n+1}^{(4)}(x) + \mathbf{P}_{n,n}^{(4)}\Phi_n^{(4)}(x) + \mathbf{P}_{n,n-1}^{(4)}\Phi_{n-1}^{(4)}(x), \tag{3.35}$$

$$xf(x)\Psi_n^{(4)}(x) = \mathbf{R}_{n,n+1}^{(4)}\Phi_{n+1}^{(4)}(x) + \mathbf{R}_{n,n}^{(4)}\Phi_n^{(4)}(x) + \mathbf{R}_{n,n-1}^{(4)}\Phi_{n-1}^{(4)}(x), \tag{3.36}$$

$$f(x)\Phi_n^{(1)}(x) = \mathbf{P}_{n,n+1}^{(1)}\Psi_{n+1}^{(1)}(x) + \mathbf{P}_{n,n}^{(1)}\Psi_n^{(1)}(x) + \mathbf{P}_{n,n-1}^{(1)}\Psi_{n-1}^{(1)}(x), \tag{3.37}$$

$$xf(x)\Phi_n^{(1)}(x) = \mathbf{R}_{n,n+1}^{(1)}\Psi_{n+1}^{(1)}(x) + \mathbf{R}_{n,n}^{(1)}\Psi_n^{(1)}(x) + \mathbf{R}_{n,n-1}^{(1)}\Psi_{n-1}^{(1)}(x), \tag{3.38}$$

where $\Phi_n^{(\beta)}(x)$ and $\Psi_n^{(\beta)}(x)$ are given in (3.18) and $\mathbf{P}_{j,k}^{(\beta)}$ and $\mathbf{R}_{j,k}^{(\beta)}$ are 2×2 quaternions. Equations (3.35)–(3.38) can be proved directly using the skew-orthogonal relation (3.20). We leave it as an exercise. In this section, we give an alternative proof by showing that the semi-infinite matrices $P^{(\beta)}$ and $R^{(\beta)}$ are tridiagonal (in the quaternion sense) and anti-self-dual.

In terms of the elements of the quaternion matrices, (3.35) and (3.36) can be written as:

$$f(x)\begin{pmatrix}\psi^{(4)}_{2n}(x)\\ \psi^{(4)}_{2n+1}(x)\end{pmatrix}=\begin{pmatrix}0 & 0\\ P^{(4)}_{2n+1,2n+2} & 0\end{pmatrix}\begin{pmatrix}\phi^{(4)}_{2n+2}(x)\\ \phi^{(4)}_{2n+3}(x)\end{pmatrix}+\begin{pmatrix}P^{(4)}_{2n,2n} & P^{(4)}_{2n,2n+1}\\ P^{(4)}_{2n+1,2n} & P^{(4)}_{2n+1,2n+1}\end{pmatrix}\begin{pmatrix}\phi^{(4)}_{2n}(x)\\ \phi^{(4)}_{2n+1}(x)\end{pmatrix}+\begin{pmatrix}0 & 0\\ P^{(4)}_{2n+1,2n-2} & 0\end{pmatrix}\begin{pmatrix}\phi^{(4)}_{2n-2}(x)\\ \phi^{(4)}_{2n-1}(x)\end{pmatrix}\tag{3.39}$$

and

$$xf(x)\begin{pmatrix}\psi^{(4)}_{2n}(x)\\ \psi^{(4)}_{2n+1}(x)\end{pmatrix}=\begin{pmatrix}R^{(4)}_{2n,2n+2} & 0\\ R^{(4)}_{2n+1,2n+2} & R^{(4)}_{2n+1,2n+3}\end{pmatrix}\begin{pmatrix}\phi^{(4)}_{2n+2}(x)\\ \phi^{(4)}_{2n+3}(x)\end{pmatrix}+\begin{pmatrix}R^{(4)}_{2n,2n} & R^{(4)}_{2n,2n+1}\\ R^{(4)}_{2n+1,2n} & R^{(4)}_{2n+1,2n+1}\end{pmatrix}\begin{pmatrix}\phi^{(4)}_{2n}(x)\\ \phi^{(4)}_{2n+1}(x)\end{pmatrix}+\begin{pmatrix}R^{(4)}_{2n,2n-2} & 0\\ R^{(4)}_{2n+1,2n-2} & R^{(4)}_{2n+1,2n-1}\end{pmatrix}\begin{pmatrix}\phi^{(4)}_{2n-2}(x)\\ \phi^{(4)}_{2n-1}(x)\end{pmatrix}.\tag{3.40}$$

For $\beta=1$, we get similar relations, where $\Phi^{(4)}$ and $\Psi^{(4)}$ are replaced by $\Psi^{(1)}$ and $\Phi^{(1)}$ respectively.

For weight functions defined in (3.23), (3.24) and (3.25), the semi-infinite matrices $P^{(\beta)}$ and $R^{(\beta)}$ have d quaternion bands above and below the diagonal [36], which is the same as saying that they satisfy $(2d+1)$-term recursion relations. Thus skew-orthogonal functions, defined with respect to the classical weights, namely the $d=1$ version of (3.23), (3.24) and (3.25)), satisfy three-term recursion relations in quaternion space.

Note. Here, we would like to mention that unlike orthogonal polynomials, the Jacobi matrix $Q^{(\beta)}$ coming from the relation $x\Phi^{(\beta)}(x)=Q^{(\beta)}\Phi^{(\beta)}(x)$, $\beta=1,4$, holds little importance in our derivation of the generalized Christoffel–Darboux formula since they do not have finite bands below the diagonal. This will be discussed further in Chapter 9.

PROOF. To prove that the skew-orthogonal functions corresponding to weight functions (3.23), (3.24) and (3.25) satisfy $(2d+1)$-term recursion relations in the quaternion space, we first show that the matrices $P^{(4)}$ and $R^{(4)}$ are anti-self-dual. For $\beta=4$, we use the scalar products

$$\sum_j P^{(4)}_{n,j}Z_{j,m}=\left(f(x)\psi^{(4)}_n(x),\psi^{(4)}_m(x)\right)=\sum_j P^{(4)}_{m,j}Z_{j,n},\tag{3.41}$$

$$\sum_j R^{(4)}_{n,j}Z_{j,m}=\left(xf(x)\psi^{(4)}_n(x),\psi^{(4)}_m(x)\right)=\sum_j R^{(4)}_{m,j}Z_{j,n}.\tag{3.42}$$

Replacing $\psi^{(4)}(x)$ by $\phi^{(1)}(x)$ for $\beta=1$, we have

$$\sum_j P^{(1)}_{n,j}Z_{m,j}=\left(f(x)\phi^{(1)}_n(x),\phi^{(1)}_m(x)\right)=\sum_j P^{(1)}_{m,j}Z_{n,j},\tag{3.43}$$

$$\sum_j R^{(1)}_{n,j} Z_{m,j} = \left(xf(x)\phi^{(1)}_n(x), \phi^{(1)}_m(x)\right) = \sum_j R^{(1)}_{m,j} Z_{n,j}. \tag{3.44}$$

Thus, we get

$$P^{(\beta)} Z = Z^{\mathrm{t}} P^{(\beta)\mathrm{t}} = -Z P^{(\beta)\mathrm{t}} \implies P^{(\beta)} = Z P^{(\beta)\mathrm{t}} Z. \tag{3.45}$$

Following the same steps for $R^{(\beta)}$, we finally get

$$\boxed{P^{(\beta)} = -P^{(\beta)\mathrm{D}}, \quad R^{(\beta)} = -R^{(\beta)\mathrm{D}}, \qquad \beta = 1, 4,} \tag{3.46}$$

where the dual of a matrix is defined as

$$A^{\mathrm{D}} := -Z A^{\mathrm{t}} Z. \tag{3.47}$$

It is straightforward to see that $P^{(\beta)}$ and $R^{(\beta)}$ have d quaternion bands (one in the case of skew-orthogonal functions defined with respect to the classical weight functions) above the diagonal. Equation (3.46) ensures that they have the same number of bands (where each entry is a 2×2 quaternion) below the diagonal. This completes the proof. □

Exercise. Verify the above results for classical and quartic weight.

4.2. Generalized Christoffel – Darboux sum. The tridiagonal symmetric Jacobi matrix $Q^{(2)}$ is instrumental in deriving the Christoffel – Darboux formula for $\beta = 2$. The symmetry (i.e., $Q^{(2)} = Q^{(2)\mathrm{t}}$) enables us to calculate the sum, while the tridiagonal property simplifies the result.

To calculate $S^{(\beta)}_{2N}(x, y)$, $\beta = 1$, 4, we need something similar to happen. In this section, we show that to evaluate $S^{(\beta)}_{2N}(x, y)$, the anti-self-dual matrix $\overline{R}^{(\beta)}(x) = R^{(\beta)} - xP^{(\beta)}$ plays a similar role for $\beta = 1, 4$ as the symmetric Jacobi matrix for $\beta = 2$. As these matrices are dependent on the weight function (or more specifically, on d), the generalized Christoffel – Darboux sum also reflects that property.

We obtain the generalized Christoffel – Darboux sum for weight functions defined in (3.23) – (3.25). To illustrate our derivation, we will refer to the classical weights, namely the Jacobi, the associated Laguerre and the Gaussian weights.

With $f(y)$ introduced to remove singularity in $w'_{\alpha_i}(x)/w_{\alpha_i}(x)$, (for example see (3.26) – (3.28) for general weight and (3.32) – (3.34) for classical weights), we use the definition of $S^{(4)}_{2N}(x, y)$ from (3.12) and the recursion relation from (3.30) to get

$$\begin{aligned}
&f(y) S^{(4)}_{2N}(x, y) - f(x) S^{(4)}_{2N}(y, x) \\
&\quad = f(y)\left[\Phi^{(4)\mathrm{t}}(x) \prod_{2N} Z \prod_{2N} \Psi^{(4)}(y)\right] + f(x)\left[\Psi^{(4)\mathrm{t}}(x) \prod_{2N} Z \prod_{2N} \Phi^{(4)}(y)\right] \\
&\quad = \left[-\Phi^{(4)\mathrm{t}}(x) Z Z \prod_{2N} Z \prod_{2N} P^{(4)} \Phi^{(4)}(y)\right] \\
&\qquad\qquad + \left[\Phi^{(4)\mathrm{t}}(x) Z Z P^{(4)\mathrm{t}} Z Z \prod_{2N} Z \prod_{2N} \Phi^{(4)}(y)\right] \\
&\quad = \left[-\widehat{\Phi}^{(4)}(x) \prod_{2N} P^{(4)} \Phi^{(4)}(y)\right] + \left[\widehat{\Phi}^{(4)}(x) P^{(4)} \prod_{2N} \Phi^{(4)}(y)\right] \\
&\quad = \widehat{\Phi}^{(4)}(x) \left[P^{(4)}, \prod_{2N}\right] \Phi^{(4)}(y).
\end{aligned} \tag{3.48}$$

Similarly,

$$\begin{aligned}(3.49)\quad & yf(y)S_{2N}^{(4)}(x,y) - xf(x)S_{2N}^{(4)}(y,x)\\ &= yf(y)\Big[\Phi^{(4)\mathrm{t}}(x)\prod_{2N} Z\prod_{2N}\Psi^{(4)}(y)\Big] + xf(x)\Big[\Psi^{(4)\mathrm{t}}(x)\prod_{2N} Z\prod_{2N}\Phi^{(4)}(y)\Big]\\ &= \Big[-\Phi^{(4)\mathrm{t}}(x)ZZ\prod_{2N} Z\prod_{2N} R^{(4)}\Phi^{(4)}(y)\Big]\\ &\qquad + \Big[\Phi^{(4)\mathrm{t}}(x)ZZR^{(4)\mathrm{t}}ZZ\prod_{2N} Z\prod_{2N}\Phi^{(4)}(y)\Big]\\ &= \Big[-\widehat{\Phi}^{(4)}(x)\prod_{2N} R^{(4)}\Phi^{(4)}(y)\Big] + \Big[\widehat{\Phi}^{(4)}(x)R^{(4)}\prod_{2N}\Phi^{(4)}(y)\Big]\\ &= \widehat{\Phi}^{(4)}(x)\Big[R^{(4)},\prod_{2N}\Big]\Phi^{(4)}(y).\end{aligned}$$

Combining the two, the generalized Christoffel – Darboux formula for symplectic ensembles of random matrices with weight functions given in (3.23) – (3.25) is given by

$$(3.50)\qquad \boxed{S_{2N}^{(4)}(x,y) = \frac{\widehat{\Phi}^{(4)}(x)[\overline{R}^{(4)}(x),\Pi_{2N}]\Phi^{(4)}(y)}{f(y)(y-x)}, \qquad N\geq d.}$$

Here, we note that for the entire family of classical weights, the above result is true with $d=1$.

For the corresponding orthogonal ensembles ($\beta=1$), the generalized Christoffel – Darboux formula is derived using similar technique. Using the recursion relation (3.31) in our definition of $S_{2N}^{(1)}(x,y)$ (3.12), we get

$$\begin{aligned}(3.51)\quad & f(x)S_{2N}^{(1)}(x,y) - f(y)S_{2N}^{(1)}(y,x)\\ &= f(x)\Big[\Phi^{(1)\mathrm{t}}(x)\prod_{2N} Z\prod_{2N}\Psi^{(1)}(y)\Big] + f(y)\Big[\Psi^{(1)\mathrm{t}}(x)\prod_{2N} Z\prod_{2N}\Phi^{(1)}(y)\Big]\\ &= \widehat{\Psi}^{(1)}(x)\Big[P^{(1)},\prod_{2N}\Big]\Psi^{(1)}(y)\end{aligned}$$

and

$$\begin{aligned}(3.52)\quad & xf(x)S_{2N}^{(1)}(x,y) - yf(y)S_{2N}^{(1)}(y,x)\\ &= xf(x)\Big[\Phi^{(1)\mathrm{t}}(x)\prod_{2N} Z\prod_{2N}\Psi^{(1)}(y)\Big] + yf(y)\Big[\Psi^{(1)\mathrm{t}}(x)\prod_{2N} Z\prod_{2N}\Phi^{(1)}(y)\Big]\\ &= \widehat{\Psi}^{(1)}(x)\Big[R^{(1)},\prod_{2N}\Big]\Psi^{(1)}(y).\end{aligned}$$

Combining the two, the generalized Christoffel – Darboux formula for orthogonal ensembles is given by

$$(3.53)\qquad \boxed{S_{2N}^{(1)}(x,y) = \frac{\widehat{\Psi}^{(1)}(x)[\overline{R}^{(1)}(y),\Pi_{2N}]\Psi^{(1)}(y)}{f(x)(x-y)}, \qquad N\geq d.}$$

Here $[\overline{R}^{(\beta)}(x)]/f(y)$, where

$$(3.54)\qquad \overline{R}^{(\beta)}(x) = R^{(\beta)} - xP^{(\beta)}, \qquad \beta=1,4,$$

is the skew-orthogonal analog of the Jacobi matrix for orthogonal polynomials.

For example, the generalized Christoffel – Darboux matrix for the Jacobi symplectic ensemble (including the associated Laguerre and Gaussian symplectic ensemble) has the following structure for $\beta = 4$:

$$
\begin{aligned}
(3.55)\quad & \widehat{\Phi}^{(4)}(x)\Big[\overline{R}^{(4)}(x), \underset{2N}{\Pi}\Big]\Phi^{(4)}(y) \\
&= \begin{pmatrix} \phi_1^{(4)}(x) \\ -\phi_0^{(4)}(x) \\ \vdots \end{pmatrix}^{\mathrm{t}}
\begin{pmatrix}
0 & 0 & 0 & 0 & 0 & 0 \\
0 & 0 & 0 & \vdots & \vdots & 0 \\
0 & 0 & 0 & -R^{(4)}_{2N-2,2N} & 0 & 0 \\
0 & 0 & 0 & -\overline{R}^{(4)}_{2N-1,2N}(x) & -R^{(4)}_{2N-1,2N+1} & 0 \\
0 & -R^{(4)}_{2N-1,2N+1} & 0 & 0 & 0 & 0 \\
0 & \overline{R}^{(4)}_{2N-1,2N}(x) & -R^{(4)}_{2N-2,2N} & 0 & \cdots & 0 \\
0 & 0 & 0 & 0 & 0 & 0 \\
0 & 0 & 0 & \vdots & \vdots & 0
\end{pmatrix}
\begin{pmatrix} \phi_0^{(4)}(y) \\ \phi_1^{(4)}(y) \\ \vdots \end{pmatrix} \\
&= R^{(4)}_{2N-2,2N}\big[\phi^{(4)}_{2N}(x)\phi^{(4)}_{2N-1}(y) - \phi^{(4)}_{2N}(y)\phi^{(4)}_{2N-1}(x)\big] \\
&\qquad + R^{(4)}_{2N-1,2N+1}\big[\phi^{(4)}_{2N-2}(x)\phi^{(4)}_{2N+1}(y) - \phi^{(4)}_{2N-2}(y)\phi^{(4)}_{2N+1}(x)\big] \\
&\qquad + \big(R^{(4)}_{2N-1,2N} - xP^{(4)}_{2N-1,2N}\big)\big[\phi^{(4)}_{2N-2}(x)\phi^{(4)}_{2N}(y) - \phi^{(4)}_{2N-2}(y)\phi^{(4)}_{2N}(x)\big].
\end{aligned}
$$

Similarly, for the family of the Jacobi orthogonal ensemble, the generalized Christoffel – Darboux matrix has the following structure for $\beta = 1$:

$$
\begin{aligned}
(3.56)\quad & \widehat{\Psi}^{(1)}(x)\Big[\overline{R}^{(1)}(y), \underset{2N}{\Pi}\Big]\Psi^{(1)}(y) \\
&= R^{(1)}_{2N-2,2N}\big[\psi^{(1)}_{2N}(x)\psi^{(1)}_{2N-1}(y) - \psi^{(1)}_{2N}(y)\psi^{(1)}_{2N-1}(x)\big] \\
&\qquad + R^{(1)}_{2N-1,2N+1}\big[\psi^{(1)}_{2N-2}(x)\psi^{(1)}_{2N+1}(y) - \psi^{(1)}_{2N-2}(y)\psi^{(1)}_{2N+1}(x)\big] \\
&\qquad + \big(R^{(1)}_{2N-1,2N} - yP^{(1)}_{2N-1,2N}\big)\big[\psi^{(1)}_{2N-2}(x)\psi^{(1)}_{2N}(y) - \psi^{(1)}_{2N-2}(y)\psi^{(1)}_{2N}(x)\big].
\end{aligned}
$$

For general d with $N \geq d$, the generalized Christoffel – Darboux matrix $\big[\overline{R}^{(\beta)}(x), \Pi_{2N}\big]$, in terms of the quaternion elements, takes the form

$$
\begin{aligned}
(3.57)\quad & \Big[\overline{\mathbf{R}}^{(\beta)}(x), \underset{N}{\Pi}\Big] \\
&= \begin{pmatrix}
0 & 0 & 0 & -\overline{\mathbf{R}}^{(\beta)}_{N-d,N}(x) & 0 & 0 \\
0 & 0 & 0 & \vdots & \ddots & 0 \\
0 & 0 & 0 & -\overline{\mathbf{R}}^{(\beta)}_{N-1,N}(x) & \cdots & -\overline{\mathbf{R}}^{(\beta)}_{N-1,N+d-1}(x) \\
\overline{\mathbf{R}}^{(\beta)}_{N,N-d}(x) & \cdots & \overline{\mathbf{R}}^{(\beta)}_{N,N-1}(x) & 0 & 0 & 0 \\
0 & \ddots & \vdots & \vdots & 0 & 0 \\
0 & 0 & \overline{\mathbf{R}}^{(\beta)}_{N+d-1,N-1}(x) & 0 & \cdots & 0
\end{pmatrix}, \\
&\hspace{25em} N \geq d.
\end{aligned}
$$

where the weight functions are defined in (3.23) – (3.25). Note that in the above matrix, every entry is a 2×2 quaternion, including those of the projection operator $\Pi_N = \operatorname{diag}(\underbrace{\mathbf{1}, \ldots, \mathbf{1}}_{N}, 0, \ldots, 0)$.

Exercise. (1) From (3.57), show that for quartic weight (i.e., $d = 2$), $S_{2N}^{(\beta)}(x,y)$ for $\beta = 1$ is given by

$$
\begin{aligned}
(x-y)S_{2N}^{(1)}(x,y) = {} & R_{2N-4,2N}^{(1)}[\psi_{2N-3}^{(1)}(x)\psi_{2N}^{(1)}(y) - (x \leftrightarrow y)] \\
& + R_{2N-2,2N+2}^{(1)}[\psi_{2N-1}^{(1)}(x)\psi_{2N+2}^{(1)}(y) - (x \leftrightarrow y)] \\
& - R_{2N-3,2N+1}^{(1)}[\psi_{2N-4}^{(1)}(x)\psi_{2N+1}^{(1)}(y) - (x \leftrightarrow y)] \\
& - R_{2N-1,2N+3}^{(1)}[\psi_{2N-2}^{(1)}(x)\psi_{2N+3}^{(1)}(y) - (x \leftrightarrow y)] \\
& + R_{2N-2,2N}^{(1)}[\psi_{2N-1}^{(1)}(x)\psi_{2N}^{(1)}(y) - (x \leftrightarrow y)] \\
& - yP_{2N-3,2N}^{(1)}[\psi_{2N-4}^{(1)}(y)\psi_{2N}^{(1)}(x) - (x \leftrightarrow y)] \\
& - \big[R_{2N-1,2N+1}^{(1)}[\psi_{2N-2}^{(1)}(x)\psi_{2N+1}^{(1)}(y) - (x \leftrightarrow y)] \\
& \quad - yP_{2N-1,2N+2}^{(1)}[\psi_{2N-2}^{(1)}(y)\psi_{2N+2}^{(1)}(x) - (x \leftrightarrow y)]\big] \\
& - yP_{2N-2,2N+1}^{(1)}[\psi_{2N-1}^{(1)}(x)\psi_{2N+1}^{(1)}(y) - (x \leftrightarrow y)] \\
& + yP_{2N-1,2N}^{(1)}[\psi_{2N-2}^{(1)}(x)\psi_{2N}^{(1)}(y) - (x \leftrightarrow y)].
\end{aligned} \tag{3.58}
$$

(2) For $\beta = 4$ and quartic weight, show

$$
\begin{aligned}
(y-x)S_{2N}^{(4)}(x,y) = {} & R_{2N-4,2N}^{(4)}[\phi_{2N-3}^{(4)}(x)\phi_{2N}^{(4)}(y) - (x \leftrightarrow y)] \\
& + R_{2N-2,2N+2}^{(4)}[\phi_{2N-1}^{(4)}(x)\phi_{2N+2}^{(4)}(y) - (x \leftrightarrow y)] \\
& - R_{2N-3,2N+1}^{(4)}[\phi_{2N-4}^{(4)}(x)\phi_{2N+1}^{(4)}(y) - (x \leftrightarrow y)] \\
& - R_{2N-1,2N+3}^{(4)}[\phi_{2N-2}^{(4)}(x)\phi_{2N+3}^{(4)}(y) - (x \leftrightarrow y)] \\
& + R_{2N-2,2N}^{(4)}[\phi_{2N-1}^{(4)}(x)\phi_{2N}^{(4)}(y) - (x \leftrightarrow y)] \\
& - xP_{2N-3,2N}^{(4)}[\phi_{2N-4}^{(4)}(y)\phi_{2N}^{(4)}(x) - (x \leftrightarrow y)] \\
& - [R_{2N-1,2N+1}^{(4)}[\phi_{2N-2}^{(4)}(x)\phi_{2N+1}^{(4)}(y) - (x \leftrightarrow y)] \\
& - xP_{2N-1,2N+2}^{(4)}[\phi_{2N-2}^{(4)}(y)\phi_{2N+2}^{(4)}(x) - (x \leftrightarrow y)]] \\
& - xP_{2N-2,2N+1}^{(4)}[\phi_{2N-1}^{(4)}(x)\phi_{2N+1}^{(4)}(y) - (x \leftrightarrow y)] \\
& + xP_{2N-1,2N}^{(4)}[\phi_{2N-2}^{(4)}(x)\phi_{2N}^{(4)}(y) - (x \leftrightarrow y)].
\end{aligned} \tag{3.59}
$$

Remark. For general weight functions (3.23) – (3.25)), there are $2d-1$ quaternions in the generalized Christoffel – Darboux formula. This is different from the Christoffel – Darboux formula for $\beta = 2$, where there is only one term irrespective of the weight function with respect to which the orthogonal polynomials are defined.

CHAPTER 4

Mapping

In order to *apply* the generalized Christoffel – Darboux formulas in random matrix theory, we need to know the asymptotic behavior of the recursion coefficients $P_{n,m}^{(\beta)}$, $R_{n,m}^{(\beta)}$ and the corresponding skew-orthogonal functions $\phi_n^{(\beta)}(x)$ and $\psi_n^{(\beta)}(x)$ for $\beta = 1$ and 4.

In this chapter, we demonstrate the existence of a mapping between orthogonal functions $\phi_n^{(2)}(x)$ and their skew-orthogonal counterparts. This helps us to obtain asymptotic results of skew-orthogonal functions and their recursion coefficients $P_{n,m}^{(\beta)}$ and $R_{n,m}^{(\beta)}$ using known properties of orthogonal functions.

1. Orthogonal polynomials and some relevant formulas

Orthogonal polynomials $P_j(x)$ of order j, associated with the weight function $w(x)$ are defined as [77]:

$$\int_{\mathbb{R}} P_j(x) P_l(x) w(x)\, \mathrm{d}x = h_j \delta_{j,l}, \qquad P_j(x) = \sum_{l=0}^{j} k_l^{(j)} x^l, \ j, l \in \mathbb{N}, \tag{4.1}$$

where h_j is the normalization constant and $k_j^{(j)}$ is the leading coefficient. They satisfy three-term recursion relation,

$$xP_j(x) = Q_{j,j+1}^{(2)} P_{j+1}(x) + Q_{j,j}^{(2)} P_j(x) + Q_{j,j-1}^{(2)} P_{j-1}(x), \qquad j = 0, 1, \ldots \tag{4.2}$$

where $Q_{j,k}^{(2)}$ are elements of the non symmetric tri-diagonal matrix, arising from the three term recursion relation of the un-normalized orthogonal polynomials. Some important results for classical weight functions (1.6), (1.7), (1.8), defined in the interval $[x_1, x_2]$, are given in Table 1

TABLE 1.

	Jacobi $[-1,1]$	Associated Laguerre $[0,\infty]$	Gaussian $[-\infty,\infty]$
$P_0(x)$	1	1	1
$P_1(x)$	$\frac{1}{2}(a+b+2)x + \frac{1}{2}(a-b)$	$-x+a+1$	$2x$
(h_j)	$h_j^{a,b} = \left(\frac{2^{a+b+1}}{(2j+a+b+1)} \frac{\Gamma(j+a+1)\Gamma(j+b+1)}{\Gamma(j+1)\Gamma(j+a+b+1)}\right)$	$h_j^a = \frac{\Gamma(j+a+1)}{j!}$	$h_j = \pi^{1/2} 2^j j!$
$k_j^{(j)}$	$k_j^{a,b} = \frac{1}{2^j}\binom{2j+a+b}{j}$	$k_j^a = \frac{(-1)^j}{j!}$	2^j
$Q_{j-1,j}^{(2)}$	$\frac{2j(j+a+b)}{(2j+a+b)(2j+a+b-1)}$	$-j$	$\frac{1}{2}$
$Q_{j-1,j-1}^{(2)}$	$\frac{(b^2-a^2)}{(2j+a+b)(2j+a+b-2)}$	$2j+a-1$	0
$Q_{j-1,j-2}^{(2)}$	$\frac{2(j+a-1)(j+b-1)}{(2j+a+b-1)(2j+a+b-2)}$	$-j+a-1$	$j-1$

For a weight function with quartic potential, we have

$$\phi_j^{(2)}(x) = \frac{(x^j + \cdots)}{\sqrt{h_j}} \exp(-NV(x)), \quad V(x) = \frac{gx^4}{4} + \frac{tx^2}{2}, \qquad g > 0. \tag{4.3}$$

We know from [10] that for even weight, with $R_m = (h_m/h_{m-1})$, $R_0 = 0$, orthogonal functions satisfy

$$x\phi_j^{(2)}(x) = \sqrt{R_{j+1}}\phi_{j+1}^{(2)}(x) + \sqrt{R_j}\phi_{j-1}^{(2)}(x), \tag{4.4}$$

where the middle term of (4.2) is zero due to the even parity of the weight function. They also satisfy

$$\frac{\mathrm{d}}{\mathrm{d}x}\phi_j^{(2)}(x) = \sum_{k=j-3}^{j+3} P_{j,k}^{(2)}\phi_k^{(2)}(x), \tag{4.5}$$

$$P_{j,j+3}^{(2)} = -Ng\sqrt{R_{j+1}R_{j+2}R_{j+3}}, \qquad P_{j,j-3}^{(2)} = Ng\sqrt{R_{j-2}R_{j-1}R_j},$$

$$P_{j,j+1}^{(2)} = -[Nt\sqrt{R_{j+1}} + Ng\sqrt{R_{j+1}}(R_j + R_{j+1} + R_{j+2})], \tag{4.6}$$

$$P_{j,j-1}^{(2)} = [Nt\sqrt{R_j} + Ng\sqrt{R_j}(R_{j-1} + R_j + R_{j+1})]. \tag{4.7}$$

Exercise. Using $\phi_{2m-1}^{(2)}(0) = 0$ in (4.4), prove

$$\phi_{2m}^{(2)}(0) = (-1)^m\sqrt{\frac{R_{2m-1}\cdots R_1}{R_{2m}\cdots R_2}}\phi_0^{(2)}(0). \tag{4.8}$$

2. Mapping between skew-orthogonal and orthogonal functions

In this section, we derive relations between skew-orthogonal and orthogonal functions. We mainly concentrate on the $\beta = 1$ family of skew-orthogonal functions, with a brief discussion on $\beta = 4$, since the methods are essentially the same. These results play a pivotal role in studying the asymptotic behavior of skew-orthogonal functions and their recursion coefficients.

We start with skew-orthogonal functions corresponding to $\beta = 1$. We expand $\phi_m^{(1)}(x)$ in the basis of orthogonal polynomials $P_j^{2\alpha_i+k}(x)$, where α_i are parameters assigned to the weight function to ensure that they have finite moments. We have

$$\phi_m^{(1)}(x) = \gamma_m^{(m)}\left[f(x)w_{\alpha_i}(x)P_{m-k}^{2\alpha_i+k}(x)\right]' + \sum^{m-1}\gamma_j^{(m)}\phi_j^{(1)}(x), \qquad m \geq k, \ (') \equiv \left(\frac{\mathrm{d}}{\mathrm{d}x}\right) \tag{4.9}$$

$\Longrightarrow$

$$\begin{aligned}\psi_m^{(1)}(x) &= \gamma_m^{(m)}\left[f(x)w_{\alpha_i}(x)P_{m-k}^{2\alpha_i+k}(x)\right] + \sum^{m-1}\gamma_j^{(m)}\psi_j^{(1)}(x) \\ &= \gamma_m^{(m)}w_{\alpha_i+k}(x)P_{m-k}^{2\alpha_i+k}(x) + \sum^{m-1}\gamma_j^{(m)}\psi_j^{(1)}(x), \quad i = 1, \ldots, 2d,\end{aligned} \tag{4.10}$$

where, $[f(x)w_{\alpha_i}(x)]'/w_{\alpha_i}(x)$ is a polynomial of order $k = 2d - 1$. The second expression is obtained from the first by multiplying by $\epsilon(y - x)$ and integrating by parts. Here, we also note that for generalized weight functions defined in (3.23)–(3.25), we have

$$f(x)w_{\alpha_i}(x) = w_{\alpha_i+k}(x); \qquad w_{\alpha_i}(x) \cdot w_{\alpha_i+k}(x) = w_{2\alpha_i+k}(x). \tag{4.11}$$

For example, for the Jacobi weight, we have

$$(1-x^2)w_{a,b}(x) = w_{a+1,b+1}(x) \qquad w_{a,b}(x)w_{a+1,b+1}(x) = w_{2a+1,2b+1}(x). \tag{4.12}$$

Equations (4.9) and (4.10) are recurrence relations which can be solved by utilizing skew-orthonormality and a careful choice of the free parameters $\gamma_k^{(m)}$. In this chapter, we solve the specific cases of quartic and the classical weight functions. An alternative derivation for the associated Laguerre and Gaussian weights are given in Appendix B.

2.1. Orthogonal ensembles ($\beta = 1$). In this section, we solve the recurrence relations (4.9) and (4.10) for the classical and quartic weights and obtain simple mapping relations between the $\beta = 1$ family of skew-orthogonal functions and their $\beta = 2$ orthogonal counterparts.

Jacobi weight function. Skew-orthogonal functions defined with respect to the Jacobi weight $w_{a,b}^2(x)$ of (1.6) are best expanded in terms of Jacobi orthogonal polynomial $P_j^{2a+1,2b+1}(x)$ corresponding to the weight function $w_{2a+1,2b+1}(x)$.

For skew-orthogonal functions with Jacobi weight, $f(x) = 1 - x^2$ and $f(x) \times w'(x)/w(x)$ is a polynomial of order 1. Hence, we expand $\phi_j^{(1)}(x)$ in the basis of orthogonal functions $\left((1-x^2)w_{a,b}(x)P_{j-1}^{2a+1,2b+1}(x)\right)'$. We have

$$\phi_m^{(1)}(x) = \gamma_m^{(m)}[(1-x^2)w_{a,b}(x)P_{m-1}^{2a+1,2b+1}(x)]' + \sum^{m-1} \gamma_k^{(m)}\phi_k^{(1)}(x), \quad m \geq 1 \tag{4.13}$$

$\Longrightarrow$

$$\begin{aligned} \psi_m^{(1)}(x) &= \gamma_m^{(m)}(1-x^2)w_{a,b}(x)P_{m-1}^{2a+1,2b+1}(x) + \sum^{m-1} \gamma_k^{(m)}\psi_k^{(1)}(x) \\ &= \gamma_m^{(m)} w_{a+1,b+1}(x)P_{m-1}^{2a+1,2b+1}(x) + \sum^{m-1} \gamma_k^{(m)}\psi_k^{(1)}(x), \qquad m \geq 1. \end{aligned} \tag{4.14}$$

Here, in the last step, we use the identity (4.12). Also, as discussed in earlier sections, the expression for $\psi_m^{(1)}(x)$ is obtained from $\phi_m^{(1)}(x)$ by multiplying by $\epsilon(y-x)$ and integrating by parts.

By using the identity

$$\begin{aligned} &\frac{\mathrm{d}}{\mathrm{d}x}\{w_{a+1,b+1}(x)P_j^{2a+1,2b+1}(x)\} \\ &\qquad = w_{a,b}(x)\{A_{j+1}P_{j+1}^{2a+1,2b+1}(x) - B_{j-1}P_{j-1}^{2a+1,2b+1}(x)\}, \end{aligned} \tag{4.15}$$

in (4.13), $\phi_j^{(1)}(x)$ can also be written as

$$\begin{aligned} \phi_m^{(1)}(x) = \gamma_m^{(m)} w_{a,b}(x)\{A_m P_m^{2a+1,2b+1}(x) - B_{m-2}P_{m-2}^{2a+1,2b+1}(x)\} \\ + \sum_{k=0}^{m-1} \gamma_k^{(m)}\phi_k^{(1)}(x), \qquad m \geq 1, \end{aligned} \tag{4.16}$$

with $\phi_0^{(1)}(x)$ being $w_{a,b}(x)$. The coefficients are given by

$$\begin{aligned} A_j &= -\frac{j(j+2a+2b+2)}{(2j+2a+2b+1)}, \\ B_j &= -\frac{(j+2a+2)(j+2b+2)}{(2j+2a+2b+5)}, \qquad B_{-1} = 0. \end{aligned} \tag{4.17}$$

Here, $\gamma_j^{(j)}$ are the leading coefficients for which we choose

$$\gamma_{2m}^{(2m)} = (A_{2m})^{-1}, \qquad \gamma_{2m+1}^{(2m+1)} = 1, \tag{4.18}$$

to fix the leading coefficient in $\pi_j^{(1)}(x)$. From the orthogonality of the Jacobi polynomials we have

$$\int w_{a+1,b+1}(x) P_j^{2a+1,2b+1}(x) \phi_k^{(1)}(x)\, \mathrm{d}x = \int w_{2a+1,2b+1}(x) P_j^{2a+1,2b+1}(x) \pi_k^{(1)}(x)\, \mathrm{d}x = 0, \tag{4.19}$$

for $k = 0, \ldots, j-1$. Using this and the skew-orthogonality relation (3.20), we have

$$0 = \left(\phi_{2m}^{(1)}(x), \psi_{2m-2k+1}^{(1)}(x)\right) \implies \gamma_{2m-2k}^{(2m)} = 0 \qquad k = 2, 3, \ldots \tag{4.20}$$

and

$$0 = \left(\psi_{2m}^{(1)}(x), \phi_{2m-2k}^{(1)}(x)\right) \implies \gamma_{2m-2k+1}^{(2m)} = 0 \qquad k = 1, 2, \ldots. \tag{4.21}$$

Also

$$0 = \left(\phi_{2m-2k}^{(1)}(x), \psi_{2m+1}^{(1)}(x)\right) \implies \gamma_{2m-2k+1}^{(2m+1)} = 0 \qquad k = 1, 2, \ldots \tag{4.22}$$

and

$$0 = \left(\psi_{2m+1}^{(1)}(x), \phi_{2m-2k+1}^{(1)}(x)\right) \implies \gamma_{2m-2k}^{(2m+1)} = 0 \qquad k = 1, 2, \ldots. \tag{4.23}$$

Thus

$$\gamma_k^{(2m)} = 0, \tag{4.24}$$

for $k \neq 2m, 2m-2$, and

$$\gamma_k^{(2m+1)} = 0, \tag{4.25}$$

$k \neq 2m, 2m+1$. Moreover, since $\gamma_{2m}^{(2m+1)}$ is arbitrary (as skew-orthogonality is not affected), we choose it to be zero. Thus we have

$$\phi_{2m}^{(1)}(x) = w_{a,b}(x) \left[P_{2m}^{2a+1,2b+1}(x) - \frac{B_{2m-2}}{A_{2m}} P_{2m-2}^{2a+1,2b+1}(x) \right] + \gamma_{2m-2}^{(2m)} \phi_{2m-2}^{(1)}(x), \qquad m \geq 1 \tag{4.26}$$

$$\psi_{2m+1}^{(1)}(x) = w_{a+1,b+1}(x) P_{2m}^{2a+1,2b+1}(x), \tag{4.27}$$

with $\left(\phi_{2m}^{(1)}(x), \psi_{2m+1}^{(1)}(x)\right)$ giving

$$g_{2m}^{(1)} = h_{2m}^{2a+1,2b+1}. \tag{4.28}$$

and $\left(\phi_{2m}^{(1)}(x), \psi_{2m-1}^{(1)}(x)\right) = 0$ fixing

$$\gamma_{2m-2}^{(2m)} = \frac{B_{2m-2}}{A_{2m}}. \tag{4.29}$$

This in turn solves the recursion relation (4.26). With these, we get the final results of the normalized skew-orthogonal functions (see also [35, 40]):

$$\phi^{(1)}_{2m}(x) = (g^{(1)}_{2m})^{-1/2} w_{a,b}(x) P^{2a+1,2b+1}_{2m}(x), \tag{4.30}$$

$$\psi^{(1)}_{2m+1}(x) = (g^{(1)}_{2m})^{-1/2} w_{a+1,b+1}(x) P^{2a+1,2b+1}_{2m}(x) \tag{4.31}$$

$$\psi^{(1)}_{2m}(x) = (g^{(1)}_{2m})^{-1/2} \Big[\frac{1}{A_{2m}} w_{a+1,b+1}(x) P^{2a+1,2b+1}_{2m-1}(x) + \gamma^{(2m)}_{2m-2} (g^{(1)}_{2m-2})^{1/2} \psi^{(1)}_{2m-2}(x) \Big] \quad (m \neq 0), \tag{4.32}$$

$$\psi^{(1)}_{0}(x) = (g^{(1)}_{0})^{-1/2} \int \epsilon(x-y) w_{a,b}(y)\, \mathrm{d}y, \tag{4.33}$$

$$\phi^{(1)}_{2m+1}(x) = (g^{(1)}_{2m})^{-1/2} w_{a,b}(x) \big(A_{2m+1} P^{2a+1,2b+1}_{2m+1}(x) - B_{2m-1} P^{2a+1,2b+1}_{2m-1}(x) \big). \tag{4.34}$$

Finally, the relation (4.15) is proved from

$$\begin{aligned} \int_{-1}^{1} \mathrm{d}x\; w_{a+1,b+1}(x) P^{2a+1,2b+1}_{j}(x) \frac{\mathrm{d}}{\mathrm{d}x} \big(w_{a+1,b+1}(x) P^{2a+1,2b+1}_{k}(x) \big) \\ = - \int_{-1}^{1} \mathrm{d}x\; w_{a+1,b+1}(x) P^{2a+1,2b+1}_{k}(x) \frac{\mathrm{d}}{\mathrm{d}x} \big(w_{a+1,b+1}(x) P^{2a+1,2b+1}_{j}(x) \big), \end{aligned} \tag{4.35}$$

which is nonzero only for $|j-k| = 1$. Now, doing the integrals for $|j-k| = 1$ in terms of $h^{2a+1,2b+1}_{j}$ and $k^{2a+1,2b+1}_{j}$, we find

$$A_j = -(j+a+b+1) \frac{k^{2a+1,2b+1}_{j-1}}{k^{2a+1,2b+1}_{j}}, \qquad B_j = \frac{h^{2a+1,2b+1}_{j+1}}{h^{2a+1,2b+1}_{j}} A_{j+1}, \tag{4.36}$$

and hence (4.17).

Exercise. Using (4.32) and (4.34), show

$$\big(\psi^{(1)}_{2m}(x), \phi^{(1)}_{2m+1}(x) \big) = -\big(\phi^{(1)}_{2m}(x), \psi^{(1)}_{2m+1}(x) \big) = -1. \tag{4.37}$$

Associated Laguerre weight function. Associated Laguerre polynomials, skew-orthogonal with respect to the associated Laguerre weight $w_a^2(x)$ (1.7), are derived and given in terms of associated Laguerre orthogonal polynomial $L^{(2a+1)}_{j}(2x)$ corresponding to the weight function $w_{2a+1}(2x)$ (see [35, 40]). Note that

$$2^{2a+1} w_a(x) w_{a+1}(x) = w_{2a+1}(2x), \tag{4.38}$$

$$w_{a+1}(x) = x w_a(x), \tag{4.39}$$

the latter vanishing at $x = 0$ for all $a > -1$. Using similar procedure, the normalized skew-orthogonal functions are given by:

$$\phi^{(1)}_{2m}(x) = (g^{(1)}_{2m})^{-1/2} 2^{a+1/2} w_a(x) L^{(2a+1)}_{2m}(2x), \tag{4.40}$$

$$\psi^{(1)}_{2m+1}(x) = (g^{(1)}_{2m})^{-1/2} 2^{a+3/2} w_{a+1}(x) L^{(2a+1)}_{2m}(2x), \tag{4.41}$$

with

$$g^{(1)}_{2m} = g^{(1)}_{2m+1} = h^{(2a+1)}_{2m}, \tag{4.42}$$

$$\psi^{(1)}_{2m}(x) = (g^{(1)}_{2m})^{-1/2}\Big[\frac{2^{a+3/2}}{A^{\mathrm{L}}_{2m}} w_{a+1}(x) L^{(2a+1)}_{2m-1}(2x) + \gamma^{\mathrm{L}}_{2m-2}(g^{(1)}_{2m-2})^{1/2}\psi^{(1)}_{2m-2}(x)\Big] \qquad (m \neq 0), \tag{4.43}$$

$$\psi^{(1)}_0(x) = (g^{(1)}_0)^{-1/2} 2^{a+1/2} \int \epsilon(x-y) w_a(y)\,\mathrm{d}y, \tag{4.44}$$

$$\phi^{(1)}_{2m+1}(x) = (g^{(1)}_{2m})^{-1/2} 2^{a+1/2} w_a(x)\big(A^{\mathrm{L}}_{2m+1} L^{(2a+1)}_{2m+1}(2x) - B^{\mathrm{L}}_{2m-1} L^{(2a+1)}_{2m-1}(2x)\big), \tag{4.45}$$

where

$$\gamma^{\mathrm{L}}_j = \frac{(j+2a+2)}{(j+2)}, \tag{4.46}$$

$$A^{\mathrm{L}}_j = j, \tag{4.47}$$

$$B^{\mathrm{L}}_j = j + 2a + 2, \qquad B^{\mathrm{L}}_{-1} = 0, \tag{4.48}$$

$$\frac{B^{\mathrm{L}}_{j-1}}{A^{\mathrm{L}}_j} h^{2a+1}_{j-1} = h^{2a+1}_j, \qquad j > 1. \tag{4.49}$$

Finally, using

$$\frac{\mathrm{d}}{\mathrm{d}x}\{w_{a+1}(x) L^{(2a+1)}_j(2x)\} = \frac{1}{2} w_a(x)\{A^{\mathrm{L}}_{j+1} L^{(2a+1)}_{j+1}(2x) - B^{\mathrm{L}}_{j-1} L^{(2a+1)}_{j-1}(2x)\}, \tag{4.50}$$

we derive (4.41) and (4.43) from (4.45) and (4.40) respectively.

Exercise. Using (4.43), (4.45) and (4.49), show

$$\big(\psi^{(1)}_{2m}(x), \phi^{(1)}_{2m+1}(x)\big) = -\big(\phi^{(1)}_{2m}(x), \psi^{(1)}_{2m+1}(x)\big) = -1. \tag{4.51}$$

Gaussian weight function. For the Gaussian weight (1.8) with $\beta = 1$, Hermite polynomials $H_j(x)$ corresponding to $\big(w(x)\big)^2 = \exp(-x^2)$ are again encountered. The skew-orthogonal polynomials can be either derived as the limiting case ($a = b \to \infty$) of the Jacobi results or be read off directly from the correlation function results of [53]. We have the normalized skew-orthogonal functions:

$$\phi^{(1)}_{2m}(x) = (g^{(1)}_{2m})^{-1/2} \mathrm{e}^{-x^2/2} H_{2m}(x), \tag{4.52}$$

$$\psi^{(1)}_{2m+1}(x) = (g^{(1)}_{2m})^{-1/2} \mathrm{e}^{-x^2/2} H_{2m}(x), \tag{4.53}$$

$$g^{(1)}_{2m} = g^{(1)}_{2m+1} = h_{2m}, \tag{4.54}$$

$$\psi^{(1)}_{2m}(x) = (g^{(1)}_{2m})^{-1/2}\big[-2\mathrm{e}^{-x^2/2} H_{2m-1}(x) + 2(2m-1)(g^{(1)}_{2m-2})^{1/2}\psi^{(1)}_{2m-2}(x)\big] \quad (m \neq 0), \tag{4.55}$$

$$\psi^{(1)}_0(x) = (g^{(1)}_0)^{-1/2} \int_{-\infty}^{\infty} \epsilon(x-y)\mathrm{e}^{-y^2/2}\,\mathrm{d}y, \tag{4.56}$$

$$\phi^{(1)}_{2m+1}(x) = (g^{(1)}_{2m})^{-1/2} \mathrm{e}^{-x^2/2}\big[-\tfrac{1}{2} H_{2m+1}(x) + 2m H_{2m-1}(x)\big]. \tag{4.57}$$

Exercise. 1. Using (4.55) and (4.57), show

$$\big(\psi^{(1)}_{2m}(x), \phi^{(1)}_{2m+1}(x)\big) = -\big(\phi^{(1)}_{2m}(x), \psi^{(1)}_{2m+1}(x)\big) = -1. \tag{4.58}$$

2. Verify the above results for the associated Laguerre and Gaussian weights with those given in Appendix B, which are obtained by taking appropriate limits.

Quartic weight. For skew-orthogonal functions with quartic weight (1.9), $f(x) = 1$ and $[f(x)w(x)]'/w(x)$ is a polynomial of order 3. Hence, without loss of generality, we expand skew-normalized function $\phi_j^{(1)}(x)$ in the basis of normalized orthogonal functions $\phi_j^{(2)}(x)$ such that

$$(4.59)\quad \phi_m^{(1)}(x) = \gamma_m^{(m)}\phi'^{(2)}_{m-3}(x) + \sum^{m-1} \gamma_k^{(m)}\phi_k^{(1)}(x) \implies \psi_m^{(1)}(x) = \gamma_m^{(m)}\phi_{m-3}^{(2)}(x) + \sum^{m-1} \gamma_k^{(m)}\psi_k^{(1)}(x).$$

for $m \geq 3$. The second expression is obtained from the first by multiplying by $\epsilon(y-x)$ and integrating by parts. Using skew-orthonormality (1.2) we get

$$(4.60)\qquad 0 = \left(\phi_{2m-2k}^{(1)}(x), \psi_{2m+1}^{(1)}(x)\right) \implies \gamma_{2m-2k+1}^{(2m+1)} = 0, \qquad k = 2, 3, \ldots$$

and

$$(4.61)\qquad 0 = \left(\psi_{2m}^{(1)}(x), \phi_{2m-2k-1}^{(1)}(x)\right) \implies \gamma_{2m-2k-2}^{(2m)} = 0, \qquad k = 2, 3, \ldots.$$

Also $\left(\psi_{2j}^{(1)}(x), \phi_{2k}^{(1)}(x)\right) = \left(\psi_{2j+1}^{(1)}(x), \phi_{2k+1}^{(1)}(x)\right) = 0$ makes $\gamma_{2k+1}^{(2m)} = \gamma_{2k}^{(2m+1)} = 0$ for $k = 0, \ldots, m-1$.

We thus get

$$(4.62)\quad \psi_{2m+1}^{(1)}(x) = \gamma_{2m+1}^{(2m+1)}\phi_{2m-2}^{(2)}(x) + \gamma_{2m-1}^{(2m+1)}\psi_{2m-1}^{(1)}(x), \qquad m > 1,$$

$$(4.63)\quad \psi_{2m}^{(1)}(x) = \gamma_{2m}^{(2m)}\phi_{2m-3}^{(2)}(x) + \gamma_{2m-2}^{(2m)}\psi_{2m-2}^{(1)}(x) + \gamma_{2m-4}^{(2m)}\psi_{2m-4}^{(1)}(x), \qquad m > 1,$$

$$(4.64)\quad \phi_{2m+1}^{(1)}(x) = \gamma_{2m+1}^{(2m+1)}\left[P_{2m-2,2m+1}^{(2)}\phi_{2m+1}^{(2)}(x) + \ldots P_{2m-2,2m-5}^{(2)}\phi_{2m-5}^{(2)}(x)\right] + \gamma_{2m-1}^{(2m+1)}\phi_{2m-1}^{(1)}(x),$$

$$(4.65)\quad \phi_{2m}^{(1)}(x) = \gamma_{2m}^{(2m)}\left[P_{2m-3,2m}^{(2)}\phi_{2m}^{(2)}(x) + \ldots P_{2m-3,2m-6}^{(2)}\phi_{2m-6}^{(2)}(x)\right] + \gamma_{2m-2}^{(2m)}\phi_{2m-2}^{(1)}(x) + \gamma_{2m-4}^{(2m)}\phi_{2m-4}^{(1)}(x).$$

The skew-orthogonality corresponding to $k=1$ along with $\left(\left(\phi_{2m-1}^{(1)}(x), \psi_{2m}^{(1)}(x)\right)=0\right)$ fixes the constants while the leading coefficients $\gamma_j^{(j)}$ is fixed by skew-normalization. We get

$$(4.66)\quad \begin{aligned} \gamma_{2m-1}^{(2m+1)} &= -\gamma_{2m+1}^{(2m+1)}\gamma_{2m-2}^{(2m-2)}P_{2m-5,2m-2}^{(2)}, \\ \gamma_{2m-4}^{(2m)} &= \gamma_{2m}^{(2m)}\gamma_{2m-3}^{(2m-3)}P_{2m-6,2m-3}^{(2)}, \\ \gamma_{2m-2}^{(2m)} &= \gamma_{2m}^{(2m)}\left[\gamma_{2m-1}^{(2m-1)}P_{2m-4,2m-3}^{(2)} + \gamma_{2m-3}^{(2m-1)}\gamma_{2m-3}^{(2m-3)}P_{2m-6,2m-3}^{(2)}\right] \end{aligned}$$

with

$$(4.67)\qquad \psi_3^{(1)}(x) = \gamma_3^{(3)}\phi_0^{(2)}(x) + \delta\psi_1^{(1)}(x),$$

$$(4.68)\qquad \phi_2^{(1)}(x) = \alpha_1\phi_2^{(2)}(x) + \alpha_2\phi_0^{(2)}(x),$$

$$(4.69)\qquad \phi_0^{(1)}(x) = c\phi_0^{(2)}(x),$$

$$(4.70)\qquad \psi_k^{(1)}(x) = \int \epsilon(x-y)\phi_k^{(1)}(y)\, dy, \qquad k = 0, 1, 2.$$

$0 = (\phi_2^{(1)}(x), \psi_5^{(1)}(x))$ gives $\alpha_1 = -\gamma_3^{(5)}$; $1 = \left(\phi_2^{(1)}(x), \psi_3^{(1)}(x)\right)$ gives $\alpha_2 = 1$ and $0 = \left(\phi_0^{(1)}(x), \psi_3^{(1)}(x)\right)$ gives $\delta = -c$.

We choose $\gamma_{2m+1}^{(2m+1)} = 1$ and $\gamma_{2m-1}^{(2m+1)} = -\sqrt{R_{2m-2}/R_{2m-3}}$. This fixes $\gamma_{2m}^{(2m)}$ which in turn fixes $\gamma_{2m-2}^{(2m)}$ and $\gamma_{2m-4}^{(2m)}$. We have

$$\psi_{2m+1}^{(1)}(x) = \phi_{2m-2}^{(2)}(x) - \sqrt{\frac{R_{2m-2}}{R_{2m-3}}}\psi_{2m-1}^{(1)}(x), \qquad m > 1. \tag{4.71}$$

Multiplying by x and using the two-term recursion relation of orthogonal polynomials, we get

$$\begin{aligned} \psi_{2m+1}^{(1)}(x) &= \sqrt{R_{2m-1}}\left[\frac{\phi_{2m-1}^{(2)}(x)}{x}\right] - \frac{\psi_1^{(1)}(x)}{\phi_{2m-2}^{(2)}(0)}, \\ \phi_{2m}^{(1)}(x) &= \frac{x\phi_{2m-1}^{(2)}(x)}{\sqrt{R_{2m-1}}}, \qquad m > 1, \end{aligned} \tag{4.72}$$

where $\phi_{2m}^{(1)}(x)$ is obtained using skew-orthonormality. Here, we note that if we choose $\psi_3^{(1)}(x)$ to be proportional to $\phi_0^{(2)}(x)$, then skew-orthonormality with $\phi_0^{(1)}(x)$ will get affected (this is possible since our expansion allows skew-orthogonal functions to form a basis set only for $m \geq 2$). However, this gets rid of the extra term in the expression for $\psi_{2m+1}^{(1)}(x)$.

$\phi_{2m+1}^{(1)}(x)$ and $\psi_{2m}^{(1)}(x)$ can be obtained by taking the derivative of $\psi_{2m+1}^{(1)}(x)$ and using skew-orthonormality respectively. $\gamma_{2m-2}^{(2m-2)}$, $\gamma_{2m-2}^{(2m)}$ and $\gamma_{2m-4}^{(2m)}$ can be calculated from (4.66).

Exercise. (i) Prove: In the asymptotic regime, $\gamma_{2m-2}^{(2m)} = 0$, $\gamma_{2m}^{(2m)} = -1/(NgR\sqrt{L})$.

(ii) Using (4.62)–(4.66), prove that in the asymptotic regime, where R_j is oscillatory for j odd and even, $(\phi_{2m}^{(1)}(x), \psi_{2m+1}^{(1)}(x)) = 1$.

Here, we note that $\gamma_{2m+1}^{(2m+1)}$ and $\gamma_{2m-1}^{(2m+1)}$ has been carefully chosen in such a manner that properties of orthogonal polynomials can be used to solve the recursion relation and obtain $\psi_{2m+1}^{(1)}(x)$ from (4.62). However, this makes the skew-orthogonal polynomials nonmonic.

2.2. Symplectic ensembles ($\beta = 4$). In this section, we solve the recurrence relations (4.9) and (4.10) for the quartic two-cut weight and obtain simple mapping relations between the $\beta = 4$ family of skew-orthogonal functions and their $\beta = 2$ orthogonal counterparts. Solutions for classical weights follow similar logic as that for $\beta = 1$ and are given as an exercise. The results however are available in Chapter 11.

Quartic weight. For skew-orthogonal functions with quartic weight (1.9), we expand

$$\phi_m^{(4)}(x) = \gamma_m^{(m)}\phi_m^{(2)}(x) + \sum^{m-1}\gamma_k^{(m)}\phi_k^{(4)}(x) \tag{4.73}$$

$$\Longrightarrow \psi_m^{(4)}(x) = \gamma_m^{(m)}\frac{\mathrm{d}}{\mathrm{d}x}\phi_m^{(2)}(x) + \sum^{m-1}\gamma_k^{(m)}\psi_k^{(4)}(x).$$

Using skew-orthonormality (1.2)), we get recursion relations

$$\phi_{2m+1}^{(4)}(x) = \gamma_{2m+1}^{(2m+1)}\phi_{2m+1}^{(2)}(x) + \gamma_{2m-1}^{(2m+1)}\phi_{2m-1}^{(4)}(x) \tag{4.74}$$

$$\phi_{2m}^{(4)}(x) = \gamma_{2m}^{(2m)}\phi_{2m}^{(2)}(x) + \gamma_{2m-2}^{(2m)}\phi_{2m-2}^{(4)}(x) + \gamma_{2m-4}^{(2m)}\phi_{2m-4}^{(4)}(x) \tag{4.75}$$

$$\psi_{2m+1}^{(4)}(x) = \gamma_{2m+1}^{(2m+1)}\left[P_{2m+1,2m+4}^{(2)}\phi_{2m+4}^{(2)}(x) + \cdots + P_{2m+1,2m-2}^{(2)}\phi_{2m-2}^{(2)}(x)\right] + \gamma_{2m-1}^{(2m+1)}\psi_{2m-1}^{(4)}(x) \tag{4.76}$$

$$\psi_{2m}^{(4)}(x) = \gamma_{2m}^{(2m)}\left[P_{2m,2m+3}^{(2)}\phi_{2m+3}^{(2)}(x) + \cdots + P_{2m,2m-3}^{(2)}\phi_{2m-3}^{(2)}(x)\right] + \gamma_{2m-2}^{(2m)}\psi_{2m-2}^{(4)}(x) + \gamma_{2m-4}^{(2m)}\psi_{2m-4}^{(4)}(x), \tag{4.77}$$

where $\phi_j^{(2)}(x)$ are the orthogonal functions defined in (4.3) and (4.4).

Using skew-orthogonality $\left(\psi_{2m-2}^{(4)}(x), \phi_{2m+1}^{(4)}(x)\right) = 0$, $\left(\psi_{2m-3}^{(4)}(x), \phi_{2m}^{(4)}(x)\right) = 0$ and finally $\left(\psi_{2m-1}^{(4)}(x), \phi_{2m}^{(4)}(x)\right) = 0$ in (4.74)–(4.77), we get

$$\begin{aligned}
\gamma_{2m-1}^{(2m+1)} &= \gamma_{2m+1}^{(2m+1)}\gamma_{2m-2}^{(2m-2)}P_{2m-2,2m+1}^{(2)},\\
\gamma_{2m-4}^{(2m)} &= -\gamma_{2m}^{(2m)}\gamma_{2m-3}^{(2m-3)}P_{2m-3,2m}^{(2)},\\
\gamma_{2m-2}^{(2m)} &= -\gamma_{2m}^{(2m)}\left[\gamma_{2m-1}^{(2m-1)}P_{2m-1,2m}^{(2)} + \gamma_{2m-3}^{(2m-1)}\gamma_{2m-3}^{(2m-3)}P_{2m-3,2m}^{(2)}\right],
\end{aligned} \tag{4.78}$$

respectively. We choose $\gamma_{2m+1}^{(2m+1)} = 1$ and $\gamma_{2m-1}^{(2m+1)} = -\sqrt{R_{2m+1}/R_{2m}}$. $\gamma_{2m-2}^{(2m-2)}$, $\gamma_{2m-2}^{(2m)}$ and $\gamma_{2m-4}^{(2m)}$ can be calculated using (4.4) in (4.78). Choice of $\gamma_{2m+1}^{(2m+1)}$ and $\gamma_{2m-1}^{(2m+1)}$ is such that we can use properties of orthogonal polynomials to solve (4.74). However, this makes the skew-orthogonal polynomials nonmonic.

Using (4.4), (4.8) and (4.78) we solve (4.74). $\psi_{2m}^{(4)}(x)$ is obtained using skew-orthonormalization (1.2):

$$\begin{aligned}
\phi_{2m+1}^{(4)}(x) &= \frac{\sqrt{R_{2m+2}}}{x}\left[\phi_{2m+2}^{(2)}(x) - \phi_{2m+2}^{(2)}(0)\exp[-NV(x)]\right],\\
\psi_{2m}^{(4)}(x) &= -\frac{x\phi_{2m+2}^{(2)}(x)}{\sqrt{R_{2m+2}}}.
\end{aligned} \tag{4.79}$$

$\phi_{2m}^{(4)}(x)$ and $\psi_{2m+1}^{(4)}(x)$ can be derived by integrating and differentiating $\psi_{2m}^{(4)}(x)$ and $\phi_{2m+1}^{(4)}(x)$ respectively.

Exercise. Using the above technique, obtain mapping relations for classical weight functions. (The results are given in Chapter 11 to demonstrate duality.)

CHAPTER 5

Unitary Ensembles

To study unitary ensembles of random matrices, we need orthogonal polynomials, a subject studied for over a century by both mathematicians and physicists. No wonder that it is the most thoroughly investigated class of matrix ensembles.

In this chapter, we use results in [77] for the classical weights and [10] for the quartic weight. We prove universality using the Christoffel–Darboux formula (3.8) and the asymptotic results for orthogonal polynomials. These asymptotic results are also helpful in deriving asymptotics for skew-orthogonal functions, using the mapping relations of Chapter 4.

1. Classical weight functions

1.1. Jacobi weight. For Jacobi weight function (1.6), $P_j(x)$ are the Jacobi polynomials $P_j^{a,b}(x)$ [77] with normalization constant $h_j^{a,b}$ and the leading coefficient $k_j^{a,b}$ of $P_j^{(a,b)}(x)$ given in Table 1.

For large j, arbitrary real a, b and fixed positive number ϵ, the normalized Jacobi polynomial is given below. For

$$x = \cos\theta, \qquad \epsilon \leq \theta \leq \pi - \epsilon,$$

$$\begin{aligned}(5.1)\quad & (h_j^{a,b})^{-1/2}\big(w_{a,b}(x)\big)^{1/2} P_j^{(a,b)}(x) \\ &= \sqrt{\frac{2}{\pi\sin\theta}}\cos\left[\left(j + \frac{a+b+1}{2}\right)\theta - \left(a+\frac{1}{2}\right)\frac{\pi}{2}\right] + O(j^{-1}),\end{aligned}$$

where we have used [77] and $h_j^{a,b} \simeq 2^{(a+b)}j^{-1}$. Here, and in other asymptotic results below, terms of lower order in j for fixed θ are being ignored. Substituting (5.1) in the Christoffel–Darboux formula (3.8), we find for large N,

(5.2)

$$\begin{aligned}S_N^{(2)}(x,y) = {} & \frac{1}{\pi\Delta\theta\sin^2\theta} \\ & \times\Bigg[\cos\left[\left(N+\frac{a+b+1}{2}\right)\theta - \left(a+\frac{1}{2}\right)\frac{\pi}{2}\right] \\ & \qquad \times \cos\left[\left(N-1+\frac{a+b+1}{2}\right)(\theta+\Delta\theta) - \left(a+\frac{1}{2}\right)\frac{\pi}{2}\right] \\ & \quad - \cos\left[\left(N+\frac{a+b+1}{2}\right)(\theta+\Delta\theta) - \left(a+\frac{1}{2}\right)\frac{\pi}{2}\right] \\ & \qquad \times \cos\left[\left(N-1+\frac{a+b+1}{2}\right)\theta - \left(a+\frac{1}{2}\right)\frac{\pi}{2}\right]\Bigg]\end{aligned}$$

$$\simeq \frac{\sin(N\Delta\theta)}{\pi\Delta\theta\sin\theta}$$
$$= \frac{\sin(N(1-x^2)^{-1/2}\Delta x)}{\pi\Delta x},$$

where we have used $k_{N-1}/k_N = \frac{1}{2} + O(1/N)$ and $h_N/h_{N-1} \simeq 1 + O(1/N)$ and the asymptotic form of the Jacobi polynomial. We have also taken $y = x + \Delta x = \cos(\theta + \Delta\theta)$ with $\Delta\theta = O(1/N)$ and expanded the cosine terms having $\theta + \Delta\theta$ in the argument. The first term gets exactly canceled. With $\Delta\theta \to 0$, we have the level density

$$S_N^{(2)}(x,x) \equiv R_1^{(2)}(x) = \frac{N}{\pi\sin\theta} = \frac{N}{\pi\sqrt{1-x^2}}, \qquad |x| < 1, \tag{5.3}$$

as derived earlier (2.65), with

$$\int_{-1}^{1} S_N^{(2)}(x,x)\,\mathrm{d}x = N. \tag{5.4}$$

Moreover, in the bulk of the spectrum, we get the universal sine-kernel

$$\lim_{N\to\infty} \frac{S_N^{(2)}(x,y)}{S_N^{(2)}(x,x)} = \frac{\sin\pi r}{\pi r} = S(r), \tag{5.5}$$

where $r = \Delta x R_1(x)$. (When $\Delta\theta = O(1)$, the above limit is zero.)

Note that the final result for the scaled kernel function is independent of a, b, thereby proving universality for the entire Jacobi family.

1.2. Associated Laguerre weight. For associated Laguerre weight function (1.7), $P_j(x)$ are the associated Laguerre polynomials $L_j^{(a)}(x)$ [77] with the normalization constant h_j^a and the leading coefficient k_j^a of $L_j^{(a)}(x)$ are given in Table 1.

For large j, formula of the Plancherel–Rotach type [70, 77] for associated Laguerre polynomial is given below. For a arbitrary and real, ϵ a fixed positive number, we have for

$$x = (4j + 2a + 2)\cos^2\theta, \qquad \epsilon \le \theta \le \pi/2 - \epsilon j^{-1/2},$$

$$\begin{aligned} &\mathrm{e}^{-x/2}L_j^{(a)}(x) \\ &\quad= \frac{(-1)^j}{\sqrt{2\pi j\sin\theta\cos\theta}}\left(\frac{j}{x}\right)^{a/2}\left\{\sin\left[\left(j + \frac{a+1}{2}\right)(\sin 2\theta - 2\theta) + \frac{3\pi}{4}\right]\right\} \\ &\qquad + (jx)^{-1/2}O(1), \end{aligned} \tag{5.6}$$

where $h_j^{(a)} \simeq j^a$. Unlike the Jacobi weight, for given x, θ depends on j; for example $\theta_j - \theta_{j\mp1} \simeq \pm(2j\tan\theta_j)^{-1}$. Then with $\theta \equiv \theta_j$, we can also write

$$\begin{aligned} &\mathrm{e}^{-x/2}L_{j\mp1}^{(a)}(x) \\ &\quad= \frac{(-1)^{j-1}}{\sqrt{2\pi j\sin\theta\cos\theta}}\left(\frac{j}{x}\right)^{a/2}\left\{\sin\left[\left(j + \frac{a+1}{2}\right)(\sin 2\theta - 2\theta) \pm 2\theta + \frac{3\pi}{4}\right]\right\} \\ &\qquad + (jx)^{-1/2}O(1). \end{aligned} \tag{5.7}$$

Substituting (5.6) and (5.7) in the Christoffel–Darboux formula (3.8), we get for

$$x = 4N\cos^2\theta, \qquad y = x + \Delta x = 4N\cos^2(\theta + \Delta\theta),$$

$$
\begin{aligned}
(5.8)\quad S_N^{(2)}(x,y) &= \frac{1}{4\pi N\Delta\theta \sin^2 2\theta} \\
&\quad\times \Bigg[\sin\left[\left(N+\frac{a+1}{2}\right)(\sin 2\theta - 2\theta) + \frac{3\pi}{4}\right] \\
&\qquad \times \sin\left[\left(N+\frac{a+1}{2}\right)(\sin 2\theta - 2\theta) + 4N\Delta\theta\sin^2\theta + 2\theta + \frac{3\pi}{4}\right] \\
&\qquad - \sin\left[\left(N+\frac{a+1}{2}\right)(\sin 2\theta - 2\theta) + 4N\Delta\theta\sin^2\theta + \frac{3\pi}{4}\right] \\
&\qquad\qquad \times \sin\left[\left(N+\frac{a+1}{2}\right)(\sin 2\theta - 2\theta) + 2\theta + \frac{3\pi}{4}\right]\Bigg] \\
&= \frac{\sin(4N\Delta\theta\sin^2\theta)}{8\pi N\Delta\theta\sin\theta\cos\theta} \\
&= \frac{\sin\left(\frac{1}{2}(4N-x)^{1/2}x^{-1/2}\Delta x\right)}{\pi\Delta x}.
\end{aligned}
$$

We have used $k_{N-1}^a/k_N^a = -N + O(1)$ and $h_N^a/h_{N-1}^a = 1 + O(1/N)$ (for large N) and the asymptotic form of the associated Laguerre polynomial.

Also, the second and fourth term in the first equation have been expanded and the first term cancels exactly. With $\Delta\theta \to 0$, we have the level density

$$(5.9)\qquad S_N^{(2)}(x,x) \equiv R_1^{(2)}(x) = \frac{\tan\theta}{2\pi} = \frac{1}{2\pi}\sqrt{\frac{4N-x}{x}}, \qquad 0 < x < 4N,$$

as in (2.70) and $\beta = 2$, with

$$(5.10)\qquad \int_0^\infty S_N^{(2)}(x,x)\,\mathrm{d}x = N.$$

Defining $r = \Delta x R_1^{(2)}(x)$, we have the $S(r)$ function as in (5.5). Hence, we have proved universality of the scaled kernel function for the associated Laguerre weight.

1.3. Gaussian weight function. For the Gaussian weight function (1.8) (with $\beta = 2$), $P_j(x)$ are the Hermite polynomials [77] $H_j(x)$, with the normalization constant h_j and the leading coefficient k_j of $H_j(x)$ is given in Table 1.

For large j, formula of the Plancherel–Rotach type for Hermite polynomial is given below. For ϵ a fixed positive number, we have for

$$x = (2j+1)^{1/2}\cos\theta, \qquad \epsilon \le \theta \le \pi - \epsilon,$$

$$
\begin{aligned}
(5.11)\quad & \mathrm{e}^{-x^2/2}H_j(x) \\
&= \frac{(h_j)^{1/2}}{\sqrt{\pi\sin\theta}}\left(\frac{2}{j}\right)^{1/4}\sin\left[\left(\frac{j}{2}+\frac{1}{4}\right)(\sin 2\theta - 2\theta) + \frac{3\pi}{4}\right] + O(j^{-1}).
\end{aligned}
$$

Unlike the Jacobi weight, for given x, θ depends on j; for example $\theta_j - \theta_{j\mp1} \simeq \pm(2j\tan\theta_j)^{-1}$. Then with $\theta \equiv \theta_j$, we can also write

$$
\begin{aligned}
(5.12)\quad & \mathrm{e}^{-x^2/2}H_{j\mp1}(x) \\
&= \frac{(h_{j\mp1})^{1/2}}{\sqrt{\pi\sin\theta}}\left(\frac{2}{j}\right)^{1/4}\sin\left[\left(\frac{j}{2}+\frac{1}{4}\right)(\sin 2\theta - 2\theta) \pm \theta + \frac{3\pi}{4}\right] + O(j^{-1}),
\end{aligned}
$$

where we have used again $\theta_j - \theta_{j\mp 1} \simeq \pm(2j\tan\theta_j)^{-1}$.

Substituting (5.11) and (5.12) in the Christoffel–Darboux formula (3.8), we get for

$$x \simeq \sqrt{2N}\cos\theta, \qquad y = x + \Delta x,$$

$$\begin{aligned}
(5.13)\quad S_N^{(2)}(x,y) &= \frac{1}{\pi\sqrt{2N}\sin^2\theta\Delta\theta}\sin\left[\left(\frac{N}{2}+\frac{1}{4}\right)(\sin 2\theta - 2\theta) + \frac{3\pi}{4}\right] \\
&\qquad \times \sin\left[\left(\frac{N}{2}+\frac{1}{4}\right)(\sin 2\theta - 2\theta) + \theta + 2N\sin^2\theta\Delta\theta + \frac{3\pi}{4}\right] \\
&\qquad - \sin\left[\left(\frac{N}{2}+\frac{1}{4}\right)(\sin 2\theta - 2\theta) + 2N\sin^2\theta\Delta\theta + \frac{3\pi}{4}\right] \\
&\qquad \times \sin\left[\left(\frac{N}{2}+\frac{1}{4}\right)(\sin 2\theta - 2\theta) + \theta + \frac{3\pi}{4}\right] \\
&= \frac{\sin(2N\Delta\theta\sin^2\theta)}{\pi\sqrt{2N}\Delta\theta\sin\theta} \\
&= \frac{\sin\left((2N-x^2)^{1/2}\Delta x\right)}{\pi\Delta x}.
\end{aligned}$$

where we have used $k_{N-1}/k_N = \frac{1}{2} + O(1/N)$ and $h_N/h_{N-1} = 2N$ and the asymptotic results of the Hermite polynomial. Also, the second and fourth term in the first equation have been expanded and the first term cancels exactly. From this, it is easy to get

$$(5.14)\qquad S_N^{(2)}(x,x) \equiv R_1^{(2)}(x) = \frac{\sqrt{2N}}{\pi}\sin\theta = \frac{\sqrt{2N-x^2}}{\pi}, \qquad |x| < \sqrt{2N},$$

as in (2.87) with

$$(5.15)\qquad \int_{-\infty}^{\infty} S_N^{(2)}(x,x)\,\mathrm{d}x = N.$$

It is easy to see that the scaled kernel function indeed gives $S(r)$ as in (5.5) with $r = \Delta x R_1^{(2)}(x)$.

2. Quartic two-cut weight

This section is based on the work of Bleher and Its [10], where we use the semi-classical asymptotic results for orthogonal polynomials corresponding to the quartic double well potential to derive the level density and the universal correlation function.

We write results obtained in [10] as $m, N \to \infty$ for orthogonal polynomials. For $\delta > 0$, in the interval $x_1 + \delta < x < x_2 - \delta$, the asymptotics of the orthogonal function is given by

$$(5.16)\qquad \phi_m^{(2)}(x) = \frac{2C_m\sqrt{x}}{\sqrt{\sin\theta}}\left[\cos\left(f_m(\theta)\right) + O(N^{-1})\right],$$

where

$$f_m(\theta) = \left(\frac{m+\frac{1}{2}}{2}\right)\left(\frac{\sin(2\theta)}{2} - \theta\right) - (-1)^m\frac{\chi}{4} + \frac{\pi}{4},$$

where

$$q = \cos\theta = \frac{gx^2 + t}{2\sqrt{\lambda' g}}, \qquad \cos\chi = \frac{2\sqrt{\lambda' g} - tq}{2\sqrt{\lambda' g}q - t}, \tag{5.17}$$
$$C_m = \frac{1}{2\sqrt{\pi}}\left(\frac{g}{\lambda}\right)^{1/4}(1 + O(N^{-1})), \qquad \lambda' = \frac{m + \frac{1}{2}}{N}.$$

We also have the asymptotic results for the recursion coefficients:

$$R_{2m+1}, R_{2m} \equiv R, L + O(N^{-2}) = \frac{-t \pm \sqrt{t^2 - 4\lambda g}}{2g}, \tag{5.18}$$
$$\text{with } \lambda \equiv \frac{m}{N} = gRL,\ t + g(R + L) = 0.$$

Equation (5.18) signifies that in the asymptotic limit, R_j oscillates between two constants R and L for j odd and even respectively.

We also have

$$\phi^{(2)}_{m\pm 1}(x) = \frac{2C_m\sqrt{x}}{\sqrt{\sin\theta}}[\cos(f_m(\theta) \mp \theta/2) + O(N^{-1})] \tag{5.19}$$

Using (5.16) and (5.19) in the Christoffel–Darboux formula (3.8), we get

$$(x - y)S^{(2)}_N(x, y) = \frac{4\sqrt{R_N}C_N^2 x}{\sin\theta}\big[\cos\big(f_N(\theta)\big)\cos(f_N(\theta + \Delta\theta) + \theta/2) \tag{5.20}$$
$$- \cos(f_N(\theta) + \theta/2)\cos\big(f_N(\theta + \Delta\theta)\big)\big]$$
$$= -4\sqrt{R_N}C_N^2\left(\frac{x\sin(\theta/2)}{\sin\theta}\right)\sin(\Delta\theta f'_N(\theta)), \qquad (') \equiv \frac{\mathrm{d}}{\mathrm{d}\theta}.$$

Here, we must remember that we have used monic polynomials ($k_N = k_{N-1} = 1$) and $\sqrt{h_N/h_{N-1}} = \sqrt{R_N} \equiv Q^{(2)}_{N-1,N}$ in the Christoffel–Darboux relation.

Also, using the asymptotic results (5.16) and (5.19) in the two-term recursion relation (4.4), with $j = N - 1$ and equating the coefficient of $\sin\big(f_N(\theta)\big)$, we have

$$x\sin\left(\frac{\theta}{2}\right) = \sqrt{R_{N-1}}\sin\theta. \tag{5.21}$$

Using this along with the relation $\sqrt{R_N R_{N-1}} = \sqrt{\lambda/g}$, and C_N from (5.17), we get from (5.20)

$$S^{(2)}_N(x, y) = \frac{\sin\big(\Delta\theta f'_N(\theta)\big)}{\pi\Delta x} + O(N^{-1}) \tag{5.22}$$
$$\implies S^{(2)}_N(x, y) = \frac{\sin\big[\Delta x N\sqrt{g}x\sqrt{1 - q^2}\big]}{\pi\Delta x} + O(N^{-1}).$$

To obtain the level-density, we take the limit

$$\lim_{\Delta x\to 0}\frac{1}{N}S^{(2)}_N(x, y) = \frac{1}{N}S^{(2)}_N(x, x) = \frac{\sqrt{g}}{\pi}|x|\sqrt{1 - q^2} + O(N^{-1}) \tag{5.23}$$
$$= \frac{|x|}{\pi}\sqrt{g - \left(\frac{gx^2 + t}{2}\right)^2} + O(N^{-1}),$$

with

$$\int_{-\infty}^{\infty} S^{(2)}_N(x, x)\,\mathrm{d}x = N. \tag{5.24}$$

For the scaled correlation function $S_N^{(2)}(x,y)/S_N^{(2)}(x,x)$, we get the sine-kernel

$$\frac{S_N^{(2)}(x,y)}{S_N^{(2)}(x,x)} = \frac{\sin \pi r}{\pi r}, \qquad r = \Delta x S_N^{(2)}(x,x). \tag{5.25}$$

Note. The asymptotic results of orthogonal functions will also be needed to work out those of skew-orthogonal functions in the following chapters.

CHAPTER 6

Orthogonal Ensembles (Even Dimension)

In Chapter 2, we gave formal expressions for correlation functions of orthogonal and symplectic ensembles, involving skew-orthogonal polynomials. There are three types of skew-orthogonal polynomials corresponding to $\beta = 1$, N even, $\beta = 1$, N odd, and $\beta = 4$. Similar to the unitary ensemble, where universality followed from the Christoffel – Darboux formula and asymptotic results of orthogonal functions, universality of correlation functions for the orthogonal and symplectic ensembles follow from the generalized Christoffel – Darboux formulas and the asymptotic properties of these skew-orthogonal functions.

In this chapter, we study skew-orthogonal functions originating from ensembles of even-dimensional random matrices, invariant under orthogonal transformation, i.e., $\beta = 1$, N even. We study ensembles with classical weight and the first nontrivial polynomial potential, namely quartic double well.

1. Method of skew-orthogonal polynomials

In terms of the quaternion kernel involving skew-orthogonal polynomials,

$$Q_N^{(1)}(x,y) = \begin{pmatrix} S_N^{(1)}(x,y) & D_N^{(1)}(x,y) \\ I_N^{(1)}(x,y) - \epsilon(x-y) & S_N^{(1)\dagger}(x,y) \end{pmatrix}, \tag{6.1}$$

where $S_N^{(1)}(x,y)$, $D_N^{(1)}(x,y)$ and $I_N^{(1)}(x,y)$ are defined by (1.22), the correlation function for $\beta = 1$, N even, is given by

$$R_n^{(1)}(x_1,\ldots,x_n) = \mathrm{Qdet}[Q_N^{(1)}(x_j,x_k)]_{j,k=1,\ldots,n}, \tag{6.2}$$

with Qdet defined in the second step of (2.47).

Since $D_N^{(1)}(x,x) = I_N^{(1)}(x,x) = \epsilon(x,x) = 0$, the level density is given by

$$R_1^{(1)}(x) = S_N^{(1)}(x,x). \tag{6.3}$$

For the universality, we prove that, with $(y-x)R_1^{(1)}(x) = r$,

$$\lim_{N\to\infty} \frac{S_N^{(1)}(x,y)}{S_N^{(1)}(x,x)} = S(r). \tag{6.4}$$

We also get

$$\lim_{N\to\infty} \frac{D_N^{(1)}(x,y)}{\left(S_N^{(1)}(x,x)\right)^2} = D(r), \tag{6.5}$$

and

$$\lim_{N\to\infty} I_N^{(1)}(x,y) = I(r), \tag{6.6}$$

thereby giving the unfolded correlation function $R_n^{(1)}$ of (2.47), (2.49). Note that (6.3) and (6.4) are also valid when $S_N^{(1)}$ is replaced by $S_N^{(1)\dagger}$. Thus we will be concerned with (6.3) and (6.4) only.

Note. In the entire chapter, we will be studying the statistical properties of ensembles of $2N \times 2N$ random matrices.

2. Classical weight functions: $\beta = 1$

In this section, we study skew-orthogonal functions arising from the classical orthogonal ensembles. We obtain asymptotic results for the skew-orthogonal functions. Using them in the generalized Christoffel – Darboux formula, we derive level densities and the universal sine-kernel in the bulk of the spectrum.

2.1. Jacobi weight: $\beta = 1$. Using the asymptotic results of Jacobi orthogonal function (5.1) in the mapping relations (4.30) – (4.34), obtained in Chapter 4, we obtain asymptotic results for the corresponding skew-orthogonal functions. Also from (4.17), we get $A_m \approx B_m \approx -m/2$ and from (4.29), we get $\gamma_{2m-2}^{(2m)} \approx 1$ for large m. Thus, for

$$x = \cos\theta, \qquad \epsilon \le \theta \le \pi - \epsilon,$$

$$\psi_{2m+1}^{(1)}(x) = \sqrt{\frac{2\sin\theta}{\pi}} \cos\left[\left(2m+a+b+\frac{3}{2}\right)\theta - \left(2a+\frac{3}{2}\right)\frac{\pi}{2}\right] + O(2m)^{-1}, \tag{6.7}$$

$$\psi_{2m}^{(1)}(x) = -\frac{1}{m\sqrt{2\pi\sin\theta}} \left[\sin\left[\left(2m+a+b+\frac{3}{2}\right)\theta - \left(2a+\frac{3}{2}\right)\frac{\pi}{2}\right] + O(1) + O(2m)^{-1}\right], \tag{6.8}$$

$$\phi_{2m+1}^{(1)}(x) = 2m\sqrt{\frac{2}{\pi\sin\theta}} \left[\sin\left[\left(2m+a+b+\frac{3}{2}\right)\theta - \left(2a+\frac{3}{2}\right)\frac{\pi}{2}\right] + O(2m)^{-1}\right], \tag{6.9}$$

$$\phi_{2m}^{(1)}(x) = \sqrt{\frac{2}{\pi\sin^3\theta}} \cos\left[\left(2m+a+b+\frac{3}{2}\right)\theta - \left(2a+\frac{3}{2}\right)\frac{\pi}{2}\right] + O(2m)^{-1}. \tag{6.10}$$

Equations (6.7), (6.9) and (6.10) are obtained by using the asymptotic relations of Jacobi orthogonal functions (5.1) into the mapping relations (4.31), (4.34) and (4.30) respectively. Equation (6.8) is obtained by partial integration of (6.10). Here, we note that by directly differentiating $\psi^{(1)}(x)$, we can get $\phi^{(1)}(x)$ to the leading order, thereby confirming our result.

Universality in Jacobi orthogonal ensemble. In this section, we calculate the level density of the Jacobi orthogonal ensemble and show that in the bulk of the spectrum, the scaled or "unfolded" [53] kernel function $S_{2N}^{(1)}(x,y)/S_{2N}^{(1)}(x,x)$ is universal.

To study the kernel function (3.56), we calculate $P_{2N-1,2N}^{(1)}$, $R_{2N-1,2N+1}^{(1)}$, $R_{2N-1,2N}^{(1)}$, $R_{2N-2,2N}^{(1)}$. We expand

$$(1-x^2)\phi_{2m+1}^{(1)}(x) = \sum_{j=2m-2}^{2m+2} P_{2m+1,j}^{(1)}\psi_j^{(1)}(x), \tag{6.11}$$

$$x(1-x^2)\phi^{(1)}_{2m+1} = \sum_{j=2m-2}^{2m+3} R^{(1)}_{2m+1,j}\psi^{(1)}_j(x), \tag{6.12}$$

$$x(1-x^2)\phi^{(1)}_{2m}(x) = \sum_{2m-2}^{2m+2} R^{(1)}_{2m,j}\psi^{(1)}_j(x). \tag{6.13}$$

Using (4.30) – (4.34) in (6.11) – (6.13), we get after some algebra:

$$P^{(1)}_{2m+1,2m+2} = \sqrt{\frac{g^{(1)}_{2m+2}}{g^{(1)}_{2m}}\frac{(2m+1)(2m+2)(2m+2a+2b+3)}{(4m+2a+2b+3)(4m+2a+2b+5)}} \times (2m+2a+2b+4), \tag{6.14}$$

$$R^{(1)}_{2m+1,2m+3} = -2\sqrt{\frac{g^{(1)}_{2m+2}}{g^{(1)}_{2m}}\frac{(2m+1)(2m+2)(2m+2a+2b+3)}{(4m+2a+2b+3)(4m+2a+2b+5)}} \times \frac{(2m+2a+2b+4)}{(4m+2a+2b+6)}, \tag{6.15}$$

$$\begin{aligned} R^{(1)}_{2m+1,2m+2} &= \sqrt{\frac{g^{(1)}_{2m+2}}{g^{(1)}_{2m}}}\frac{[(2b+1)^2-(2a+1)^2](2m+1)(2m+2)}{(4m+2a+2b+3)(4m+2a+2b+4)} \times \frac{(2m+2a+2b+3)(2m+2a+2b+4)}{(4m+2a+2b+5)(4m+2a+2b+6)}, \\ &= 0, \qquad \text{for } a=b, \end{aligned} \tag{6.16}$$

$$R^{(1)}_{2m,2m+2} = -2\sqrt{\frac{g^{(1)}_{2m+2}}{g^{(1)}_{2m}}\frac{(2m+1)(2m+2)(2m+2a+2b+3)}{(4m+2a+2b+3)(4m+2a+2b+4)}} \times \frac{(2m+2a+2b+4)}{(4m+2a+2b+5)}. \tag{6.17}$$

For $m = N-1$, large N, we have $g^{(1)}_{2N}/g^{(1)}_{2N-2} \simeq 1$. This gives us for large N

$$\begin{aligned} P^{(1)}_{2N-1,2N} &\sim N^2 + O(N), \qquad & R^{(1)}_{2N-1,2N+1} &\sim -\frac{N}{2} + O(1), \\ R^{(1)}_{2N-1,2N} &\sim (b-a)O(1), \qquad & R^{(1)}_{2N-2,2N} &\sim -\frac{N}{2} + O(1). \end{aligned} \tag{6.18}$$

Finally, using (6.18) and the asymptotic results for $\psi^{(1)}_j(x)$, for both j odd (6.7) and even (6.8), defined in the θ interval $[\epsilon, \pi-\epsilon]$, in the generalized Christoffel–Darboux formula (3.56), we get for

$$x = \cos\theta, \qquad y = x + \Delta x = \cos(\theta + \Delta\theta), \qquad x - y \simeq \Delta\theta \sin\theta,$$

$$\begin{aligned} &(x-y)(1-x^2)S^{(1)}_{2N}(x,y) \\ &= R^{(1)}_{2N-2,2N}\left[\psi^{(1)}_{2N}(x)\psi^{(1)}_{2N-1}(y) - \psi^{(1)}_{2N}(y)\psi^{(1)}_{2N-1}(x)\right] \\ &\quad + R^{(1)}_{2N-1,2N+1}\left[\psi^{(1)}_{2N-2}(x)\psi^{(1)}_{2N+1}(y) - \psi^{(1)}_{2N-2}(y)\psi^{(1)}_{2N+1}(x)\right] \\ &\quad + \left(R^{(1)}_{2N-1,2N} - yP^{(1)}_{2N-1,2N}\right)\left[\psi^{(1)}_{2N-2}(x)\psi^{(1)}_{2N}(y) - \psi^{(1)}_{2N-2}(y)\psi^{(1)}_{2N}(x)\right] \end{aligned}$$

$$
\begin{aligned}
&= \frac{1}{2\pi}\left[\sin(f_{2N}(\theta))\cos(f_{2N-2}(\theta+\Delta\theta)) - \sin(f_{2N}(\theta+\Delta\theta))\cos(f_{2N-2}(\theta))\right] \\
&\quad + \frac{1}{2\pi}\left[\sin(f_{2N-2}(\theta))\cos(f_{2N}(\theta+\Delta\theta)) - \sin(f_{2N-2}(\theta+\Delta\theta))\cos(f_{2N}(\theta))\right] \\
&\quad - \frac{\cos\theta}{2\pi\sin\theta}\left[\sin(f_{2N-2}(\theta))\sin(f_{2N}(\theta+\Delta\theta)) - \sin(f_{2N-2}(\theta+\Delta\theta))\sin(f_{2N}(\theta))\right],
\end{aligned}
$$

where, in the second step, we have dropped $O(1/N^2)$ term. Thus we get

$$
\begin{aligned}
(6.19)\quad & (x-y)(1-x^2)S_{2N}^{(1)}(x,y) \\
&= \frac{1}{2\pi}\left[\sin[f_{2N}(\theta) - f_{2N-2}(\theta+\Delta\theta)] + \sin[f_{2N-2}(\theta) - f_{2N}(\theta+\Delta\theta)]\right] \\
&\qquad\qquad + \frac{\cos\theta\sin(2\theta)}{2\pi\sin\theta}\left[\sin[\Delta\theta f'_{2N}(\theta)]\right] \\
&= -\frac{1}{2\pi}\left[\sin[\Delta\theta f'_{2N}(\theta) - 2\theta] + \sin[\Delta\theta f'_{2N}(\theta) + 2\theta]\right] \\
&\qquad\qquad + \frac{\cos\theta\sin(2\theta)}{2\pi\sin\theta}\left[\sin[\Delta\theta f'_{2N}(\theta)]\right] \\
&\simeq -\frac{1}{\pi}\sin[2N\Delta\theta]\left[\cos(2\theta) - \cos^2\theta\right]
\end{aligned}
$$

which gives us finally

$$
(6.20)\qquad S_{2N}^{(1)}(x,y) = \frac{\sin(2N\Delta\theta)}{\pi\Delta\theta\sin\theta} = \frac{\sin(2N(1-x^2)^{-1/2}\Delta x)}{\pi\Delta x}, \qquad |x| < 1.
$$

With $x \to y$, we get the level density $\rho(x)$,

$$
(6.21)\qquad S_{2N}^{(1)}(x,x) := R_1(x) = \frac{2N}{\pi\sqrt{1-x^2}}, \qquad |x| < 1.
$$

with

$$
(6.22)\qquad \int_{-1}^{1} S_{2N}^{(1)}(x,x)\,\mathrm{d}x = 2N.
$$

With $\Delta x \to 0$, i.e., in the bulk of the spectrum, the unfolded kernel function

$$
(6.23)\qquad \frac{S_{2N}^{(1)}(x,y)}{S_{2N}^{(1)}(x,x)} = \frac{\sin\pi r}{\pi r}, \qquad r = R_1(x)\Delta x,
$$

is independent of $w_{a,b}(x)$ and is the universal sine-kernel.

2.2. Associated Laguerre weight: $\beta = 1$. To obtain the asymptotic formula for the associated Laguerre skew-orthogonal functions, we use asymptotic results of the corresponding orthogonal functions (5.6) in the mapping relations of Chapter 4. To get $\phi_{2m}^{(1)}(x)$ (4.40) and $\psi_{2m+1}^{(1)}(x)$ (4.41), we use the relation $2^{a+1/2}\sqrt{x}w_a(x) = \sqrt{w_{2a+1}(2x)}$ and $2^{a+1/2}w_{a+1}(x) = \sqrt{x}\sqrt{w_{2a+1}(2x)}$. Replacing this in (4.40) and (4.41) respectively and using (5.6) we get the asymptotic formula.

For arbitrary a and ϵ a fixed positive number,

$$
2x = (8m+4a+4)\cos^2\theta, \qquad \epsilon \le \theta \le \pi/2 - \epsilon m^{-1/2}, \qquad \theta \equiv \theta_{2m},
$$

$$
(6.24)\qquad \phi_{2m}^{(1)}(x) = \frac{1}{4m\sqrt{\pi}\sin\theta\cos^3\theta}\left[\sin(f_{2m}(\theta)) + \frac{O(1)}{\sqrt{2mx}}\right],
$$

$$\psi^{(1)}_{2m+1}(x) = \frac{2}{\sqrt{\pi \tan\theta}}\left[\sin(f_{2m}(\theta)) + \frac{O(1)}{\sqrt{2mx}}\right], \tag{6.25}$$

$$\psi^{(1)}_{2m}(x) = -\frac{1}{4m\sqrt{\pi \sin^3\theta\cos\theta}}\left[\cos(f_{2m}(\theta)) + O(1) + \frac{O(1)}{\sqrt{2mx}}\right], \tag{6.26}$$

$$\phi^{(1)}_{2m+1}(x) = 2\sqrt{\frac{\tan\theta}{\pi}}\left[\cos(f_{2m}(\theta)) + \frac{O(1)}{\sqrt{2mx}}\right], \tag{6.27}$$

with

$$\phi^{(1)}_{2m\pm 2}(x) = \frac{1}{4m\sqrt{\pi \sin\theta\cos^3\theta}}\left[\sin(f_{2m}(\theta) \mp 4\theta) + \frac{O(1)}{\sqrt{2mx}}\right]. \tag{6.28}$$

Similar relations hold for (6.25), (6.26) and (6.27) as $m \to m \pm 1$. Here

$$f_{2m}(\theta) = (2m + a + 1)(\sin 2\theta - 2\theta) + \frac{3\pi}{4}. \tag{6.29}$$

Universality in associated Laguerre orthogonal ensemble. In this section, we calculate the level density of the associated Laguerre orthogonal ensemble and show that in the bulk of the spectrum, the scaled or "unfolded" [53] kernel function $S^{(1)}_{2N}(x,y)/S^{(1)}_{2N}(x,x)$ is universal.

To calculate the kernel function (3.56), we need $R^{(1)}_{2N-1,2N+1}$, $R^{(1)}_{2N-1,2N}$, $R^{(1)}_{2N-2,2N}$ and $P^{(1)}_{2N-1,2N}$. For this, we expand

$$2x\phi^{(1)}_{2m+1}(x) = \sum_{j=2m-2}^{2m+2} P^{(1)}_{2m+1,j}\psi^{(1)}_j(x), \tag{6.30}$$

$$2x^2\phi^{(1)}_{2m}(x) = \sum_{j=2m-2}^{2m+2} R^{(1)}_{2m,j}\psi^{(1)}_j(x), \tag{6.31}$$

$$2x^2\phi^{(1)}_{2m+1}(x) = \sum_{j=2m-2}^{2m+3} R^{(1)}_{2m+1,j}\psi^{(1)}_j(x), \tag{6.32}$$

Using (4.45) in (6.30), we get

$$P^{(1)}_{2m+1,2m+2} = \sqrt{\frac{g^{(1)}_{2m+2}}{g^{(1)}_{2m}}}(2m+1)(2m+2). \tag{6.33}$$

Using (4.40) and (4.45) in (6.31) and (6.32), we get

$$R^{(1)}_{2m,2m+2} = R^{(1)}_{2m+1,2m+3} = -\sqrt{\frac{g^{(1)}_{2m+2}}{g^{(1)}_{2m}}}(m+1)(2m+1), \tag{6.34}$$

and

$$R^{(1)}_{2m+1,2m+2} = \sqrt{\frac{g^{(1)}_{2m+2}}{g^{(1)}_{2m}}}(m+1)(2m+1)(4m+2a+4). \tag{6.35}$$

For $m = N-1$, large N, we have $g^{(1)}_{2N}/g^{(1)}_{2N-2} \simeq 1$. This gives us for large N

$$\begin{aligned} P^{(1)}_{2N-1,2N} &\sim 4N^2 + O(N), & R^{(1)}_{2N-1,2N+1} &\sim -2N^2 + O(N), \\ R^{(1)}_{2N-1,2N} &\sim 8N^3 + O(N^2), & R^{(1)}_{2N-2,2N} &\sim -2N^2 + O(N). \end{aligned} \tag{6.36}$$

Finally, using (6.25), (6.26) and (6.28), defined in the θ interval $[\epsilon, \pi/2 - \epsilon N^{-1/2}]$, and (6.36) in the generalized Christoffel–Darboux formula (3.56), we get for

$$x = (4N + 2a + 2)\cos^2\theta, \qquad y = x + \Delta x,$$

$$\begin{aligned}(6.37)\quad & 2x(x-y)S_{2N}^{(1)}(x,y)\\ &= R_{2N-2,2N}^{(1)}\left[\psi_{2N}^{(1)}(x)\psi_{2N-1}^{(1)}(y) - \psi_{2N}^{(1)}(y)\psi_{2N-1}^{(1)}(x)\right]\\ &\quad + R_{2N-1,2N+1}^{(1)}\left[\psi_{2N-2}^{(1)}(x)\psi_{2N+1}^{(1)}(y) - \psi_{2N-2}^{(1)}(y)\psi_{2N+1}^{(1)}(x)\right]\\ &\quad + (R_{2N-1,2N}^{(1)} - yP_{2N-1,2N}^{(1)})\left[\psi_{2N-2}^{(1)}(x)\psi_{2N}^{(1)}(y) - \psi_{2N-2}^{(1)}(y)\psi_{2N}^{(1)}(x)\right]\\ &= \frac{N}{\pi\sin^2\theta}\left[\sin\big(f_{2N-2}(\theta+\Delta\theta) - f_{2N}(\theta)\big) + \sin\big(f_{2N}(\theta+\Delta\theta) - f_{2N-2}(\theta)\big)\right]\\ &\quad - \left[\frac{N}{\pi\sin^2\theta\tan 2\theta}\right]\left[\cos\big(f_{2N-2}(\theta)\big)\cos\big(f_{2N}(\theta+\Delta\theta)\big)\right.\\ &\qquad\qquad \left.- \cos\big(f_{2N-2}(\theta+\Delta\theta)\big)\cos\big(f_{2N}(\theta)\big)\right]\\ &= \frac{2N\cos 4\theta}{\pi\sin^2\theta}\sin\left(\Delta\theta\frac{\partial f_{2N}(\theta)}{\partial\theta}\right) - \left[\frac{N\sin 4\theta}{\pi\sin^2\theta\tan 2\theta}\right]\sin\left(\Delta\theta\frac{\partial f_{2N}(\theta)}{\partial\theta}\right).\end{aligned}$$

Here, we have neglected the oscillatory term of $O(1)$ arising from the even function. Finally, we get

$$\begin{aligned}(6.38)\qquad (x-y)S_{2N}^{(1)}(x,y) &= \frac{1}{\pi\sin^2 2\theta}(\cos 4\theta - \cos^2 2\theta)\left[\sin\left(\Delta\theta\frac{\partial f_{2N}(\theta)}{\partial\theta}\right)\right]\\ &= \frac{1}{\pi}\sin[8N\sin^2\theta\Delta\theta].\end{aligned}$$

Combining all, we get

$$(6.39)\qquad S_{2N}^{(1)}(x,y) = \frac{\sin(x^{-1/2}\sqrt{(4N-x)}\Delta x)}{\pi\Delta x}, \qquad 0 < x < 4N.$$

With $x \to y$, we get the level density

$$(6.40)\qquad S_{2N}^{(1)}(x,x) = \frac{1}{\pi}\sqrt{\frac{4N-x}{x}}, \qquad 0 < x < 4N.$$

with

$$(6.41)\qquad \int_0^\infty S_{2N}^{(1)}(x,x)\,\mathrm{d}x = 2N.$$

With $\Delta x \to 0$, (i.e., in the bulk of the spectrum) and $r = \Delta x S_{2N}^{(1)}(x,x)$, the unfolded kernel function

$$(6.42)\qquad \frac{S_{2N}^{(1)}(x,y)}{S_{2N}^{(1)}(x,x)} = \frac{\sin\pi r}{\pi r},$$

is independent of $w_a(x)$ and is the universal sine-kernel.

2.3. Gaussian weight: $\beta = 1$. To obtain the asymptotic formula for Hermite skew-orthogonal functions, we use asymptotic results of the corresponding orthogonal functions (5.11) in the mapping relations of Chapter 4. To get $\phi_{2m}^{(1)}(x)$ and $\psi_{2m+1}^{(1)}(x)$, we use the asymptotic results of orthogonal functions (5.11) in (4.52) and (4.53) respectively. The same is true for the other pair of skew-orthogonal functions.

Using $x = (4m+1)^{1/2}\cos\theta$, we get (5.11) for $\phi^{(1)}_{2m}(x)$ and $\psi^{(1)}_{2m+1}(x)$ with $j = 2m$, while for $\psi^{(1)}_{2m}(x)$ and $\phi^{(1)}_{2m+1}(x)$ we have

$$\psi^{(1)}_{2m}(x) = -\frac{1}{2m^{3/4}\sqrt{\pi}\sin^3\theta}\cos\left[\left(m+\frac{1}{4}\right)(\sin 2\theta - 2\theta) + \frac{3\pi}{4}\right], \tag{6.43}$$

$$\phi^{(1)}_{2m+1}(x) = 2m^{1/4}\sqrt{\frac{\sin\theta}{\pi}}\cos\left[\left(m+\frac{1}{4}\right)(\sin 2\theta - 2\theta) + \frac{3\pi}{4}\right]. \tag{6.44}$$

Universality in GOE. In this section, we calculate the level density of Gaussian orthogonal ensemble and show that in the bulk of the spectrum, the scaled or "unfolded" [53] kernel function $S^{(1)}_{2N}(x,y)/S^{(1)}_{2N}(x,x)$ is universal.

To calculate the kernel function (3.56), we need $R^{(1)}_{2N-1,2N+1}$, $R^{(1)}_{2N-1,2N}$, $R^{(1)}_{2N-2,2N}$ and $P^{(1)}_{2N-1,2N}$. For this, we expand

$$2\phi^{(1)}_{2m+1}(x) = \sum_{j=2m-2}^{2m+2} P^{(1)}_{2m+1,j}\psi^{(1)}_j(x), \tag{6.45}$$

$$2x\phi^{(1)}_{2m}(x) = \sum_{j=2m-2}^{2m+2} R^{(1)}_{2m,j}\psi^{(1)}_j(x), \tag{6.46}$$

$$2x\phi^{(1)}_{2m+1}(x) = \sum_{j=2m-2}^{2m+3} R^{(1)}_{2m+1,j}\psi^{(1)}_j(x), \tag{6.47}$$

Using (4.57) in (6.45), we get

$$P^{(1)}_{2m+1,2m+2} = \frac{1}{4}\sqrt{\frac{g^{(1)}_{2m+2}}{g^{(1)}_{2m}}}. \tag{6.48}$$

Using (4.52) and (4.57) in (6.46) and (6.47), we get

$$R^{(1)}_{2m,2m+2} = R^{(1)}_{2m+1,2m+3} = -\frac{1}{4}\sqrt{\frac{g^{(1)}_{2m+2}}{g^{(1)}_{2m}}}, \tag{6.49}$$

and

$$R^{(1)}_{2m+1,2m+2} = 0. \tag{6.50}$$

For $m = N-1$, large N, we have from (4.54) and Table 1, $g^{(1)}_{2N}/g^{(1)}_{2N-2} \simeq 4N$. This gives us for large N,

$$\begin{gathered} P^{(1)}_{2N-1,2N} \sim N + O(1), \\ R^{(1)}_{2N-1,2N+1} \sim -N + O(1), \qquad R^{(1)}_{2N-2,2N} \sim -N + O(1). \end{gathered} \tag{6.51}$$

Finally using (6.43) and the integral of (6.44), defined in the θ interval $[\epsilon, \pi-\epsilon]$, and (6.51) in the generalized Christoffel–Darboux formula (3.56), we get for

$$x = (4N+1)^{1/2}\cos\theta, \qquad y = x + \Delta x,$$

$$
\begin{aligned}
(6.52)\quad & (x-y)S_{2N}^{(1)}(x,y) \\
&= R_{2N-2,2N}^{(1)}\left[\psi_{2N}^{(1)}(x)\psi_{2N-1}^{(1)}(y)-\psi_{2N}^{(1)}(y)\psi_{2N-1}^{(1)}(x)\right] \\
&\quad + R_{2N-1,2N+1}^{(1)}\left[\psi_{2N-2}^{(1)}(x)\psi_{2N+1}^{(1)}(y)-\psi_{2N-2}^{(1)}(y)\psi_{2N+1}^{(1)}(x)\right] \\
&\quad - yP_{2N-1,2N}^{(1)}\left[\psi_{2N-2}^{(1)}(x)\psi_{2N}^{(1)}(y)-\psi_{2N-2}^{(1)}(y)\psi_{2N}^{(1)}(x)\right] \\
&= \frac{1}{2\pi\sin^2\theta}\left[\sin\left(f_{2N-2}(\theta+\Delta\theta)-f_{2N}(\theta)\right)+\sin\left(f_{2N}(\theta+\Delta\theta)-f_{2N-2}(\theta)\right)\right] \\
&\quad - \left[\frac{\cos\theta}{2\pi\sin^3\theta}\right]\left[\cos\left(f_{2N-2}(\theta)\right)\cos\left(f_{2N}(\theta+\Delta\theta)\right)\right. \\
&\quad \left. -\cos\left(f_{2N-2}(\theta+\Delta\theta)\right)\cos\left(f_{2N}(\theta)\right)\right] \\
&= \frac{\cos 2\theta}{\pi\sin^2\theta}\sin\left(\Delta\theta\frac{\partial f_{2N}(\theta)}{\partial\theta}\right)-\left[\frac{\cos\theta\sin 2\theta}{2\pi\sin^3\theta}\right]\sin\left(\Delta\theta\frac{\partial f_{2N}(\theta)}{\partial\theta}\right).
\end{aligned}
$$

Here, we have neglected the oscillatory term of $O(1)$ arising from the even function. Finally, we get

$$
\begin{aligned}
(6.53)\qquad (x-y)S_{2N}^{(1)}(x,y) &= \frac{1}{\pi\sin^2\theta}(\cos 2\theta-\cos^2\theta)\left[\sin\left(\Delta\theta\frac{\partial f_{2N}(\theta)}{\partial\theta}\right)\right] \\
&= \frac{1}{\pi}\sin[4N\sin^2\theta\Delta\theta].
\end{aligned}
$$

Combining all, we get

$$
(6.54)\qquad S_{2N}^{(1)}(x,y)=\frac{\sin(\sqrt{(4N-x^2)}\Delta x)}{\pi\Delta x},\qquad |x|<\sqrt{4N}.
$$

With $x\to y$, we get the semi-circular level density

$$
(6.55)\qquad S_{2N}^{(1)}(x,x)=\frac{1}{\pi}\sqrt{4N-x^2},\qquad |x|<\sqrt{4N}.
$$

with

$$
(6.56)\qquad \int_{-\infty}^{\infty}S_{2N}^{(1)}(x,x)\,\mathrm{d}x=2N.
$$

With $\Delta x\to 0$, (i.e., in the bulk of the spectrum) and $r=\Delta x S_{2N}^{(1)}(x,x)$, the unfolded kernel function

$$
(6.57)\qquad \frac{S_{2N}^{(1)}(x,y)}{S_{2N}^{(1)}(x,x)}=\frac{\sin\pi r}{\pi r},
$$

is independent of $w(x)$ and is the universal sine-kernel.

3. Quartic two-cut weight: $\beta=1$

In this section, we use results obtained in Chapters 4 and 5 to study the asymptotic properties of skew-orthogonal functions corresponding to quartic two-cut potential. Having obtained that, we use them in the generalized Christoffel – Darboux formula (3.58) to study the corresponding matrix model.

We know that the leading terms in the mapping relation is given by

$$
(6.58)\quad \psi_{2m+1}^{(1)}(x)\simeq\sqrt{R_{2m-1}}\left[\frac{\phi_{2m-1}^{(2)}(x)}{x}\right],\quad \phi_{2m}^{(1)}(x)=\frac{x\phi_{2m-1}^{(2)}(x)}{\sqrt{R_{2m-1}}},\qquad m>1.
$$

$\phi^{(1)}_{2m+1}(x)$ and $\psi^{(1)}_{2m}(x)$ can be obtained by taking derivative of $\psi^{(1)}_{2m+1}(x)$ and using skew-orthonormality, respectively.

Now, we derive asymptotics for skew-orthogonal functions using results for orthogonal functions [10]. In the limit $m, N \to \infty$, we assume $\epsilon > 0$, such that $\epsilon < \lambda(= m/N) < \lambda_{\rm cr}(= t^2/4g)$. This ensures that $\phi^{(1)}_m(x)$ and $\psi^{(1)}_m(x)$ are concentrated on two intervals $[-x_2, -x_1]$ and $[x_1, x_2]$ and are exponentially small outside. Using results from Chapter 4 for large m for orthogonal polynomials we have for $x_{1,2} = \sqrt{(-t \mp 2\sqrt{\lambda g})/g}$, in the range $x_1 + \delta < x < x_2 - \delta$, $\delta > 0$ and $m > 1$,

$$\psi^{(1)}_{2m+1}(x) \simeq 2C_{2m-1}\sqrt{\frac{R}{x\sin\theta}}\left[\cos\big(f_{2m-1}(\theta)\big) + O\left(\frac{1}{N}\right)\right], \tag{6.59}$$

$$\psi^{(1)}_{2m}(x) \simeq \frac{C_{2m-1}}{N\lambda}\sqrt{\frac{Lx}{\sin^3\theta}}\left[\sin\big(f_{2m-1}(\theta)\big) + O\left(\frac{1}{N}\right)\right], \tag{6.60}$$

$$\phi^{(1)}_{2m+1}(x) \simeq -4N\lambda C_{2m-1}\sqrt{\frac{x\sin\theta}{L}}\left[\sin\big(f_{2m-1}(\theta)\big) + O\left(\frac{1}{N}\right)\right], \tag{6.61}$$

$$\phi^{(1)}_{2m}(x) \simeq 2C_{2m-1}\sqrt{\frac{x^3}{R\sin\theta}}\left[\cos\big(f_{2m-1}(\theta)\big) + O\left(\frac{1}{N}\right)\right], \tag{6.62}$$

where

$$\begin{aligned} q = \cos\theta &= \frac{gx^2+t}{2\sqrt{\lambda' g}}, \qquad \cos\chi = \frac{2\sqrt{\lambda' g} - tq}{2\sqrt{\lambda' g}q - t}, \\ C_m &= \frac{1}{2\sqrt{\pi}}\left(\frac{g}{\lambda}\right)^{1/4}\big(1 + O(N^{-1})\big). \end{aligned} \tag{6.63}$$

Here, $\psi^{(1)}_{2m}(x)$ is calculated by expressing the asymptotic result of $\phi^{(1)}_{2m}(x)$ as a derivative and integrating by parts while $\phi^{(1)}_{2m+1}(x)$ is calculated from $\psi^{(1)}_{2m+1}(x)$ by simply taking its derivative. We note that, for a given x, θ varies with m. We have chosen $\theta \equiv \theta_{2m-1}$ such that

$$\psi^{(1)}_{2m+1\pm k}(x) \simeq 2C_{2m-1}\sqrt{\frac{R}{x\sin\theta}}\big[\cos\big(f_{2m-1}(\theta) \mp (k\theta)/2\big) + O(1/N)\big], \qquad m \gg k, \tag{6.64}$$

and so on for $\psi^{(1)}_{2m\pm k}(x)$, $\phi^{(1)}_{2m+1\pm k}(x)$, $\phi^{(1)}_{2m\pm k}(x)$.

For the recursion coefficients $P^{(1)}_{m,n}$ and $R^{(1)}_{m,n}$, we simply read off from (4.62)–(4.65). We get for large m,

$$\begin{aligned} P^{(1)}_{2m,2m+3} &= \sqrt{\frac{L}{R}}; \qquad P^{(1)}_{2m+1,2m+4} = Ra^2; \\ P^{(1)}_{2m,2m+1} &= -\frac{t}{gR}; \qquad P^{(1)}_{2m+1,2m+2} = 0; \\ R^{(1)}_{2m,2m+4} &= R^{(1)}_{2m+1,2m+5} = -N\lambda; \\ R^{(1)}_{2m,2m+2} &= R^{(1)}_{2m+1,2m+3} = Nt\sqrt{RL}, \end{aligned} \tag{6.65}$$

where $a = -N\sqrt{g\lambda}$.

For the quartic potential, $\beta = 1$, with $f_{2N-1}(\theta) \equiv f(\theta)$, the kernel function is given by (3.58)(also see [36]). The first four terms give

$$
\begin{aligned}
-\frac{2C_{2N-1}^2\sqrt{LR}}{\sin^2\theta}&\big[\cos(f(\theta)+2\theta)\sin\big(f(\theta+\Delta\theta)\big)-\sin\big(f(\theta)\big)\cos(f(\theta+\Delta\theta)+2\theta)\\
&+\cos(f(\theta)+\theta)\sin(f(\theta+\Delta\theta)-\theta)-\cos(f(\theta+\Delta\theta)+\theta)\sin\big(f(\theta)-\theta\big)\\
&+\sin(f(\theta+\Delta\theta)+2\theta)\cos\big(f(\theta)\big)-\sin(f(\theta)+2\theta)\cos\big(f(\theta+\Delta\theta)\big)\\
&+\sin(f(\theta+\Delta\theta)+\theta)\cos(f(\theta)-\theta)-\sin(f(\theta)+\theta)\cos(f(\theta+\Delta\theta)-\theta)\big]\\
&\qquad\qquad\qquad\qquad\qquad\qquad\qquad\qquad\qquad\qquad=-\frac{2\cos 2\theta\sin\big(\Delta\theta f'(\theta)\big)}{\pi\sin^2\theta}
\end{aligned}
$$

Following similar algebra, the next four terms give

$$
\begin{aligned}
\frac{4C_{2N-1}^2 t\cos\theta\sin(\Delta\theta f'_{2N-1}(\theta))}{g\sin^2\theta}&+\frac{4x^2 gC_{2N-1}^2\cos\theta\sin(\Delta\theta f'_{2N-1}(\theta))}{g\sin^2\theta}\\
&=\frac{4C_{2N-1}^2\cos\theta\sin(\Delta\theta f'_{2N-1}(\theta))}{g\sin^2\theta}\left[\frac{gx^2+t}{2\sqrt{\lambda' g}}\right]2\sqrt{\lambda' g}\\
&=\frac{2\cos^2\theta\sin(\Delta\theta f'_{2N-1}(\theta))}{\pi\sin^2\theta}
\end{aligned}
\tag{6.66}
$$

while the last term gives $-\sin(\Delta\theta f'_{2N-1}(\theta))/\pi$. Here, we have used the asymptotic results (as $m, N \to \infty$):

$$
\begin{gathered}
\cos\theta=\frac{gx^2+t}{2\sqrt{\lambda' g}},\qquad C_m=\frac{1}{2\sqrt{\pi}}\left(\frac{g}{\lambda}\right)^{1/4}(1+O(N^{-1})),\\
\lambda'=\lambda+O(N^{-1}).
\end{gathered}
\tag{6.67}
$$

Combining all, we get for $x = y + \Delta y$

$$
\begin{aligned}
(x-y)S_{2N}^{(1)}(x,y)&=\frac{\sin\big(\Delta\theta f'_{2N-1}(\theta)\big)}{\pi}+O(N^{-1})\\
\implies S_{2N}^{(1)}(x,y)&=\frac{\sin\big[\Delta y 2N\sqrt{g}y\sqrt{1-q^2}\big]}{\pi\Delta y}+O(N^{-1}).
\end{aligned}
\tag{6.68}
$$

To obtain the level-density, we take the limit

$$
\begin{aligned}
\lim_{\Delta y\to 0}\frac{1}{2N}S_{2N}^{(1)}(x,y)=\frac{1}{2N}S_{2N}^{(1)}(y,y)&=\frac{\sqrt{g}}{\pi}|y|\sqrt{1-q^2}+O(N^{-1})\\
&=\frac{|y|}{\pi}\sqrt{g-\left(\frac{gy^2+t}{2}\right)^2}+O(N^{-1}),
\end{aligned}
\tag{6.69}
$$

with

$$
\int_{-\infty}^{\infty}S_{2N}^{(1)}(y,y)\,\mathrm{d}y=2N.
\tag{6.70}
$$

For the scaled correlation kernel, $S_{2N}^{(1)}(x,y)/S_{2N}^{(1)}(y,y)$ we get

$$
\frac{S_{2N}^{(1)}(x,y)}{S_{2N}^{(1)}(y,y)}=\frac{\sin\pi r}{\pi r},\qquad r=\Delta y S_{2N}^{(1)}(y,y).
\tag{6.71}
$$

Note. Orthogonal ($\beta = 1$) and symplectic ($\beta = 4$) ensembles with quartic double well potential have been studied [74–76] using the Sener – Verbaarschot – Widom formalism. Here, as well as in the following chapters, we use the method of skew-orthogonal polynomials to derive the same results.

CHAPTER 7

Orthogonal Ensembles (Odd Dimension)

When N is odd, the formal results of Chapter 4 have to be modified [27] because the last skew-orthogonal polynomial is unpaired and its normalization is left arbitrary by (1.2). In this case we supplement (1.2) by the extra condition

$$\int_{\mathbb{R}} \pi_j^{(1)}(x)w(x)\,\mathrm{d}x = \delta_{j,N-1}, \tag{7.1}$$

and introduce the additional functions

$$M(x) = \phi_{N-1}^{(1)}(x) \qquad \mu(x) = \psi_{N-1}^{(1)}(x). \tag{7.2}$$

The result (6.2) is still valid if we define the quaternion $Q_N^{(1)}$ for $\beta = 1$, N odd, by

$$Q_N^{(1)}(x,y) = \begin{pmatrix} S_{N-1}^{(1)}(x,y) + M(x) & D_{N-1}^{(1)}(x,y) \\ I_{N-1}^{(1)}(x,y) - \epsilon(x-y) + \mu(x) - \mu(y) & S_{N-1}^{(1)\dagger}(x,y) + M(y) \end{pmatrix}. \tag{7.3}$$

This kernel also satisfies the normalization and convolution properties as done by (6.1) for N even.

Because of the extra condition (7.1) skew-orthogonal polynomials $\pi_j^{(1)}(x)$ of order j cannot be constructed. However, we can construct $\pi_j^{(1)}(x)$ of order $j+1$ for $j = 0, \ldots, N-2$ and then construct $\pi_{N-1}^{(1)}(x)$ of order $N-1$; thus $\pi_{N-2}^{(1)}(x)$ and $\pi_{N-1}^{(1)}(x)$ will both be of order $(N-1)$ while there will be no polynomial of order 0.

For the Jacobi weight function (1.6), we have, with $m = 0, 1, \ldots, (N-3)/2$,

$$\phi_{2m}^{(1)}(x)w_{a,b}(x)(P_{2m+1}^{2a+1,2b+1}(x) - c_m), \tag{7.4}$$

$$c_m = \int_{\mathbb{R}} w_{a,b}(y)P_{2m+1}^{2a+1,2b+1}(y)\,\mathrm{d}y, \tag{7.5}$$

$$\psi_{2m+1}^{(1)}(x) = w_{a+1,b+1}(x)P_{2m+1}^{2a+1,2b+1}(x), \tag{7.6}$$

while

$$g_{2m}^{(1)} = g_{2m+1}^{(1)} = h_{2m+1}^{2a+1,2b+1}, \tag{7.7}$$

and

$$\phi_{N-1}^{(1)}(x) = \frac{w_{a,b}(x)P_{N-1}^{2a+1,2b+1}(x)}{\int w_{a,b}(y)P_{N-1}^{2a+1,2b+1}(y)\,\mathrm{d}y}. \tag{7.8}$$

To prove these results, note first that, without the constant c_m, (7.4), (7.6) satisfy the skew-orthogonality as in (4.30), (4.31) of the previous chapter. The constant c_m ensures that the condition (7.1) is satisfied for $\phi_{2m}^{(1)}(x)$ without affecting the skew-orthogonality with the other functions. The condition (7.1) is automatically satisfied for $\phi^{(1)}_{2m+1}(x)$, because $\psi^{(1)}_{2m+1}(1) = 0$. Finally $\phi^{(1)}_{N-1}(x)$ of (7.8) is skew-orthogonal with the other $\psi^{(1)}$'s and is normalized according to (7.1). The results for $\phi^{(1)}_{2m+1}(x)$ and $\psi^{(1)}_{2m}(x)$ can be worked out along the steps followed in the case for N even. For the asymptotic forms, (6.7) – (6.10) apply with the replacement $m \to m+\frac{1}{2}$ in the phases of the main cosine and sine terms. Note that the additional c_m term in (7.5) does not contribute to the asymptotic forms of the kernel function, nor do the additional terms involving μ and M. We have thus the same level density and the universal unfolded correlation functions as in the even-N case.

For the associated Laguerre weight, we again have (4.40) – (4.42) with $2m \to 2m+1$ on the right-hand sides and the c_m term of the type (7.4), (7.5) for $\phi^{(1)}_{2m}(x)$. For $\phi^{(1)}_{N-1}(x)$, we have (7.8) with $P^{2a+1,2b+1}_{N-1} \to L^{(2a+1)}_{N-1}$. The asymptotic forms (6.24) – (6.27) again apply with the replacement $m \to m+\frac{1}{2}$ in the phases.

For the Gaussian weight, we again have (4.52) – (4.54) with $2m \to 2m+1$ on the right-hand side. In this case the c_m-correction of the type (7.5) is not needed, as $c_m = 0$. With $P_{N-1} \to H_{N-1}$ in (7.8) we get $\phi^{(1)}_{N-1}(x)$. These results are identical with the Gaussian ensemble results of [53].

Similar arguments are valid for quartic potential as well. We again have (4.72) with $2m \to 2m+1$ on the right-hand side. In this case the c_m-correction of the type (7.5) is not needed, as $c_m = 0$. With $w(x)P_{N-1}(x) \to \phi^{(2)}_{N-1}(x)$ in (7.8) we get $\phi^{(1)}_{N-1}(x)$.

CHAPTER 8

Symplectic Ensembles

In this chapter, we study the asymptotic properties of skew-orthogonal functions and their recursion coefficients corresponding to $\beta = 4$. As in Chapters 4 and 5, we consider skew-orthogonal functions corresponding to the family of Jacobi weights (1.6) (including the limiting cases of associated Laguerre (1.7) and Gaussian (1.8) weights) as well as weight function with quartic double well potential. As an application, we use them to calculate the level densities and the universal correlations for symplectic ensembles of random matrices. Tools used are the generalized Christoffel – Darboux formula and the asymptotic results of these skew-orthogonal functions.

1. Method of skew-orthogonal polynomials

Using the kernel functions defined in (1.22) the correlation functions for $\beta = 4$ are then given by

$$Q_{2N}^{(4)}(x,y) = \begin{pmatrix} S_{2N}^{(4)}(x,y) & D_{2N}^{(4)}(x,y) \\ I_{2N}^{(4)}(x,y) & S_{2N}^{(4)\dagger}(x,y) \end{pmatrix}. \tag{8.1}$$

As in $\beta = 1$, the level density is given by

$$R_1^{(4)}(x) = S_{2N}^{(4)}(x,x). \tag{8.2}$$

Here we would like to remind the readers that the two different definitions of normalization of skew-orthogonal functions (see (2.34) and (2.35) and the discussion that follows) corresponding to $\beta = 4$ lead to different definitions of level density. We will see this when we study the quartic weight.

With $\Delta x R_1(x) = r$, we need to prove

$$\lim_{N\to\infty} \frac{S_{2N}^{(4)}(x,y)}{S_{2N}^{(4)}(x,x)} = S(2r), \tag{8.3}$$

$$\lim_{N\to\infty} \frac{D_{2N}^{(4)}(x,y)}{S_{2N}^{(4)}(x,x)} = D(2r), \tag{8.4}$$

$$\lim_{N\to\infty} \frac{I_{2N}^{(4)}(x,y)}{S_{2N}^{(4)}(x,x)} = I(2r), \tag{8.5}$$

to obtain the universal $R_n^{(4)}$ for $\beta = 4$. Since $D_{2N}^{(4)}(x,y)$ and $I_{2N}^{(4)}(x,y)$ follow from $S_{2N}^{(4)}(x,y)$, we would again be concerned only with $S_{2N}^{(4)}(x,y)$.

Note. From the duality relations in Chapter 11, we can study $\beta = 4$ through a mapping between $\phi^{(4)} \to \psi^{(1)}$ and $\psi^{(4)} \to \phi^{(1)}$, where the skew-orthogonal functions are defined with respect to $w^2(x)$ instead of $w(x)$. This is left as an exercise.

Here, we prove it using $w(x)$ to demonstrate that by doing so, you may lose the duality structure between the two ensembles but the "universality" relations remain unaffected.

2. Classical weight functions: $\beta = 4$

In this section, we study asymptotic properties of the family of Jacobi skew-orthogonal functions arising in the study of symplectic ensembles of random matrices. This also includes the limiting cases of associated Laguerre and Gaussian weight functions.

We use the asymptotic results for the skew-orthogonal functions, defined with respect to $w(x)$, (instead of $w^2(x)$) and the generalized Christoffel–Darboux formula to prove universality. Here we would like to mention that our generalized Christoffel–Darboux results are valid for both these definitions with some minor differences in $\overline{R}^{(4)}(x)$.

2.1. Jacobi weight: $\beta = 4$. We consider skew-orthogonal functions defined with respect to the Jacobi weight (1.6). As shown in [35, 40], $\pi_j^{(4)}(x)$ and $\pi_j^{(4)\prime}(x)$ can be written compactly in terms of the Jacobi orthogonal polynomials $P_j^{a,b}(x)$:

$$\pi_{2m+1}^{(4)\prime}(x) = P_{2m}^{a,b}(x), \qquad m = 0, 1, \ldots, \tag{8.6}$$

$$\pi_{2m}^{(4)\prime}(x) = P_{2m-1}^{a,b}(x) + \eta_{2m}\pi_{2m-2}^{(4)\prime}(x), \quad m = 1, 2, \ldots, \qquad \pi_0^{(4)\prime}(x) = 0, \tag{8.7}$$

where η_{2m} is a constant, given in (8.14) below. On integration, we find the polynomials

$$\pi_{2m+1}^{(4)}(x) = \frac{2}{(2m+a+b)}\left[D_{2m+1}P_{2m+1}^{a,b}(x) + E_{2m+1}P_{2m}^{a,b}(x) + F_{2m+1}P_{2m-1}^{a,b}(x)\right], \quad m = 0, 1, \ldots, \tag{8.8}$$

$$\pi_{2m}^{(4)}(x) = \frac{2}{(2m+a+b-1)}\left[D_{2m}P_{2m}^{a,b}(x) + E_{2m}P_{2m-1}^{a,b}(x) + F_{2m}P_{2m-2}^{a,b}(x)\right] + \eta_{2m}\pi_{2m-2}^{(4)}(x), \quad m = 0, 1, \ldots. \tag{8.9}$$

Here, $P_j^{a,b}(x) = 0$ for negative j. In (8.8) and (8.9) we have used the indefinite integral

$$\begin{aligned}\frac{1}{2}(j+a+b)\int P_j^{a,b}(x)\,\mathrm{d}x &= P_{j+1}^{a-1,b-1}(x) \\ &= D_{j+1}P_{j+1}^{a,b}(x) + E_{j+1}P_j^{a,b}(x) + F_{j+1}P_{j-1}^{a,b}(x).\end{aligned} \tag{8.10}$$

The integration constants have been put equal to zero because of skew-orthogonality with $\pi_1^{(4)}(x)$. The constants D_j, E_j, F_j, η_j and $g_j^{(4)}$ are given by

$$D_j = \frac{(j+a+b)(j+a+b-1)}{(2j+a+b)(2j+a+b-1)}, \tag{8.11}$$

$$E_j = \frac{(a-b)(j+a+b-1)}{(2j+a+b)(2j+a+b-2)}, \tag{8.12}$$

$$F_j = -\frac{(j+a-1)(j+b-1)}{(2j+a+b-1)(2j+a+b-2)}, \tag{8.13}$$

$$\eta_j = \frac{(j+a-1)(j+b-1)(2j+a+b-5)}{(j-1)(j+a+b-1)(2j+a+b-1)}, \tag{8.14}$$

$$(8.15)\qquad g_{2m}^{(4)} = g_{2m+1}^{(4)} = \frac{2h_{2m}^{a,b}}{4m+a+b-1} = \frac{2^{a+b+2}\Gamma(2m+a+1)\Gamma(2m+b+1)}{(4m+a+b+1)(4m+a+b-1)\Gamma(2m+1)\Gamma(2m+a+b+1)}.$$

For large j and large m,

$$(8.16)\qquad D_j = -F_j = \frac{1}{4} + O(j^{-1}), \qquad E_j = (a-b)\left[\frac{1}{4j} + O(j^{-2})\right],$$

$$(8.17)\qquad \eta_j = 1 + O(j^{-1}), \qquad g_{2m}^{(4)} = \frac{2^{a+b}}{4m^2} + O(m^{-3}),$$

and in the same approximation,

$$(8.18)\qquad \pi_{2m+1}^{(4)}(x) = \frac{1}{4m}[P_{2m+1}^{a,b}(x) - P_{2m-1}^{a,b}(x)],$$

$$(8.19)\qquad \pi_{2m}^{(4)}(x) = \frac{1}{4m}[P_{2m}^{a,b}(x) + 2^{(a+b)/2}(w_{a,b}(x))^{-1/2}].$$

Here in (8.19), the nonpolynomial term on the right-hand side is the large-m approximation for the lower-order terms in the series in (8.9) and has been verified numerically.

For large m, we use (8.18), (8.19) and the asymptotic formula for Jacobi orthogonal polynomial (5.1) to obtain asymptotic formulas for Jacobi skew-orthogonal functions. For a, b arbitrary and real, and ϵ a fixed positive number, we have for

$$x = \cos\theta, \qquad \epsilon \le \theta \le \pi - \epsilon,$$

$$(8.20)\qquad (g_{2m}^{(4)})^{-1/2}\sqrt{w_{a,b}(x)}\pi_{2m+1}^{(4)}(x) := \phi_{2m+1}^{(4)}(x) = -\sqrt{\frac{\sin\theta}{\pi m}}\sin[f_{2m}(\theta)] + O(m^{-3/2}),$$

$$(8.21)\qquad (g_{2m}^{(4)})^{-1/2}\sqrt{w_{a,b}(x)}\pi_{2m}^{(4)}(x) := \phi_{2m}^{(4)}(x) = \frac{1}{2}\left[\frac{1}{\sqrt{\pi m}\sin\theta}\cos[f_{2m}(\theta)] + 1\right] + O(m^{-3/2}),$$

$$(8.22)\qquad (g_{2m}^{(4)})^{-1/2}\sqrt{w_{a,b}(x)}\pi_{2m+1}^{(4)\prime}(x) = 2\sqrt{\frac{m}{\pi\sin\theta}}\cos[f_{2m}(\theta)] + O(m^{-1/2}),$$

$$(8.23)\qquad (g_{2m}^{(4)})^{-1/2}\sqrt{w_{a,b}(x)}\pi_{2m}^{(4)\prime}(x) = \sqrt{\frac{m}{\pi\sin^3\theta}}[\sin[f_{2m}(\theta)] + 1] + O(m^{-1/2}),$$

where

$$(8.24)\qquad f_{2m}(\theta) = \left(2m + \frac{a+b+1}{2}\right)\theta - \left(a + \frac{1}{2}\right)\frac{\pi}{2}.$$

Equation (8.22) is derived from (8.6) while (8.23) is obtained by differentiating (8.19). Here, we would like to mention that to calculate the level density and two-point correlation function for the Jacobi symplectic ensemble, (8.22) and (8.23) are used to calculate the recursion coefficients $P_{j,k}^{(4)}$ and $R_{j,k}^{(4)}$. They are also important in defining the skew-orthogonal functions.

Level density and 2-point correlation for the Jacobi symplectic ensemble. In this section, we calculate the level density of the Jacobi symplectic ensemble and show that, in the bulk of the spectrum, the scaled or "unfolded" [53] kernel function $S_{2N}^{(4)}(x,y)/S_{2N}^{(4)}(x,x)$ is universal.

From (3.55) we see that to use the generalized Christoffel–Darboux formula for classical weights, we need to calculate $R_{2N-1,2N+1}^{(4)}$, $R_{2N-1,2N}^{(4)}$, $R_{2N-2,2N}^{(4)}$ and $P_{2N-1,2N}^{(4)}$. We use $f(x) = 1 - x^2$ in the recursion relation (3.39), that is

$$(1-x^2)\frac{\mathrm{d}}{\mathrm{d}x}\phi_{2m+1}^{(4)}(x) = \sum_{j=2m-2}^{2m+2} P_{2m+1,j}^{(4)}\phi_j^{(4)}(x), \tag{8.25}$$

to get $P_{2m+1,2m+2}^{(4)}$. It is given by

$$P_{2m+1,2m+2}^{(4)} = -\sqrt{\frac{g_{2m+2}^{(4)}}{g_{2m}^{(4)}}}\,\frac{(2m+a+b+1)Q_{2m,2m+1}^{(2)}Q_{2m+1,2m+2}^{(2)}}{2D_{2m+2}}, \tag{8.26}$$

where $Q_{j,k}^{(2)}$ and D_j is given in Table 1 and (8.11) respectively.

Similarly, we use (3.40):

$$\begin{aligned} x(1-x^2)\frac{\mathrm{d}}{\mathrm{d}x}\phi_{2m}^{(4)}(x) &= \sum_{j=2m-2}^{2m+2} R_{2m,j}^{(4)}\phi_j^{(4)}(x),\\ x(1-x^2)\frac{\mathrm{d}}{\mathrm{d}x}\phi_{2m+1}^{(4)}(x) &= \sum_{j=2m-2}^{2m+3} R_{2m+1,j}^{(4)}\phi_j^{(4)}(x), \end{aligned} \tag{8.27}$$

to get $R_{2m,2m+2}^{(4)}$, $R_{2m+1,2m+3}^{(4)}$ and $R_{2m+1,2m+2}^{(4)}$. For large m, we get to the leading order

$$\begin{aligned} R_{2m+1,2m+2}^{(4)} &\simeq \sqrt{\frac{g_{2m+2}^{(4)}}{g_{2m}^{(4)}}}\Bigg[\frac{(2m+a+b+1)E_{2m+3}Q_{2m,2m+1}^{(2)}Q_{2m+1,2m+2}^{(2)}Q_{2m+2,2m+3}^{(2)}}{2D_{2m+3}D_{2m+2}}\\ &\qquad - (a-b)\frac{(2m+a+b+1)}{2(2m+a+b)}\frac{D_{2m+1}}{D_{2m+2}}Q_{2m+1,2m+2}^{(2)}\Bigg]\\ &\qquad + (a-b)O(1/m)\\ &= 0 \qquad \text{for } a=b, \end{aligned} \tag{8.28}$$

$$R_{2m+1,2m+3}^{(4)} \simeq -\sqrt{\frac{g_{2m+2}^{(4)}}{g_{2m}^{(4)}}}\,\frac{(2m+a+b+2)Q_{2m,2m+1}^{(2)}Q_{2m+1,2m+2}^{(2)}Q_{2m+2,2m+3}^{(2)}}{2D_{2m+3}} + O(1), \tag{8.29}$$

$$R_{2m,2m+2}^{(4)} \simeq -\sqrt{\frac{g_{2m+2}^{(4)}}{g_{2m}^{(4)}}}\,\frac{(2m+a+b+1)Q_{2m-1,2m}^{(2)}Q_{2m,2m+1}^{(2)}Q_{2m+1,2m+2}^{(2)}}{2D_{2m+2}} + O(1), \tag{8.30}$$

where $Q_{j,k}^{(2)}$, D_j and E_j are given in Table 1, (8.11) and (8.12) respectively.

For large j, we use $Q^{(2)}_{j,j+1} \simeq \frac{1}{2}$. Also for $m = N-1$, large N, we have $g^{(4)}_{2N}/g^{(4)}_{2N-2} \simeq 1$. This gives us for large N

$$\begin{aligned} P^{(4)}_{2N-1,2N} &\sim -N + O(1), & R^{(4)}_{2N-1,2N+1} &\sim -\frac{N}{2} + O(1), \\ R^{(4)}_{2N-1,2N} &\sim \frac{(b-a)}{4} + (b-a)O(N^{-1}), & R^{(4)}_{2N-2,2N} &\sim -\frac{N}{2} + O(1). \end{aligned} \tag{8.31}$$

Finally, using (8.20) and (8.21), defined in the θ interval $[\epsilon, \pi-\epsilon]$, and (8.31) in the generalized Christoffel–Darboux formula (3.55), we get for

$$x = \cos\theta, \qquad y = x + \Delta x = \cos(\theta + \Delta\theta), \qquad x - y \simeq \Delta\theta \sin\theta,$$

$$\begin{aligned} &(y-x)(1-y^2)S^{(4)}_{2N}(x,y) \\ &= R^{(4)}_{2N-2,2N}\left[\phi^{(4)}_{2N}(x)\phi^{(4)}_{2N-1}(y) - \phi^{(4)}_{2N}(y)\phi^{(4)}_{2N-1}(x)\right] \\ &\quad + R^{(4)}_{2N-1,2N+1}\left[\phi^{(4)}_{2N-2}(x)\phi^{(4)}_{2N+1}(y) - \phi^{(4)}_{2N-2}(y)\phi^{(4)}_{2N+1}(x)\right] \\ &\quad + \left(R^{(4)}_{2N-1,2N} - xP^{(4)}_{2N-1,2N}\right)\left[\phi^{(4)}_{2N-2}(x)\phi^{(4)}_{2N}(y) - \phi^{(4)}_{2N-2}(y)\phi^{(4)}_{2N}(x)\right] \\ &= \frac{1}{4\pi}\left[\cos\big(f_{2N}(\theta)\big)\sin\big(f_{2N-2}(\theta+\Delta\theta)\big)\right. \\ &\qquad\qquad \left. - \cos\big(f_{2N}(\theta+\Delta\theta)\big)\sin\big(f_{2N-2}(\theta)\big)\right] \\ &\quad + \frac{1}{4\pi}\left[\cos\big(f_{2N-2}(\theta)\big)\sin\big(f_{2N}(\theta+\Delta\theta)\big)\right. \\ &\qquad\qquad \left. - \cos\big(f_{2N-2}(\theta+\Delta\theta)\big)\sin\big(f_{2N}(\theta)\big)\right] \\ &\quad + \frac{\cos\theta}{4\pi\sin\theta}\left[\cos\big(f_{2N-2}(\theta)\big)\cos\big(f_{2N}(\theta+\Delta\theta)\big)\right. \\ &\qquad\qquad \left. - \cos\big(f_{2N-2}(\theta+\Delta\theta)\big)\cos\big(f_{2N}(\theta)\big)\right] \\ &= \frac{1}{4\pi}\left[\sin[f_{2N-2}(\theta+\Delta\theta) - (f_{2N}(\theta)] + \sin[f_{2N}(\theta+\Delta\theta) - f_{2N-2}(\theta)]\right] \\ &\qquad\qquad - \frac{\cos\theta\sin(2\theta)}{4\pi\sin\theta}\left[\sin[\Delta\theta f'_{2N}(\theta)]\right] \\ &= \frac{1}{4\pi}\left[\sin\left[\left(2N + \frac{a+b+1}{2}\right)\Delta\theta - 2\theta\right] + \sin\left[\left(2N + \frac{a+b+1}{2}\right)\Delta\theta + 2\theta\right]\right] \\ &\qquad\qquad - \frac{\cos\theta\sin(2\theta)}{4\pi\sin\theta}\left[\sin\left[\left(2N + \frac{a+b+1}{2}\right)\Delta\theta\right]\right] \\ &\simeq \frac{1}{2\pi}\sin[2N\Delta\theta]\left[\cos(2\theta) - \cos^2\theta\right], \end{aligned} \tag{8.32}$$

where in the second step, we have dropped the $O(N^{-1})$ terms. This gives us

$$S^{(4)}_{2N}(x,y) = \frac{\sin(2N\Delta\theta)}{2\pi\Delta\theta\sin\theta} = \frac{\sin(2N(1-x^2)^{-1/2}\Delta x)}{2\pi\Delta x}, \qquad |x| < 1. \tag{8.33}$$

With $\Delta x \to 0$, we get the level density

$$S^{(4)}_{2N}(x,x) = \frac{N}{\pi\sqrt{1-x^2}}, \qquad |x| < 1, \tag{8.34}$$

with

$$\int_{-1}^{1} S_{2N}^{(4)}(x,x)\,\mathrm{d}x = N. \tag{8.35}$$

In the bulk of the spectrum, the unfolded kernel function is independent of $w_{a,b}(x)$ and gives the universal sine-kernel

$$\frac{S_{2N}^{(4)}(x,y)}{S_{2N}^{(4)}(x,x)} = \frac{\sin 2\pi r}{2\pi r}, \tag{8.36}$$

where $r = \Delta x S_{2N}^{(4)}(x,x)$.

2.2. Associated Laguerre weight: $\beta = 4$. Now we consider the associated Laguerre skew-orthogonal functions (1.7) and study their asymptotic properties. As shown in [35, 40], $\pi_j^{(4)}(x)$ and $\pi_j^{(4)\prime}(x)$ can be written compactly in terms of the associated Laguerre orthogonal polynomials $L_j^{(a)}(x)$:

$$\pi_{2m+1}^{(4)\prime}(x) = L_{2m}^{(a)}(x), \qquad m = 0, 1, \ldots, \tag{8.37}$$

$$\pi_{2m}^{(4)\prime}(x) = L_{2m-1}^{(a)}(x) + \binom{2m+a-1}{2m-1}\pi_{2m-2}^{(4)\prime}(x), \qquad m = 1, 2, \ldots, \tag{8.38}$$

$$\pi_0^{(4)\prime}(x) = 0.$$

On integration, we find

$$\pi_{2m+1}^{(4)}(x) = -L_{2m+1}^{(a)}(x) + L_{2m}^{(a)}(x), \qquad m = 0, 1, \ldots, \tag{8.39}$$

$$\pi_{2m}^{(4)}(x) = -L_{2m}^{(a)}(x) + L_{2m-1}^{(a)}(x) + \binom{2m+a-1}{2m-1}\pi_{2m-2}^{(4)}, \quad m = 0, 1, \ldots. \tag{8.40}$$

For $a = 0$, (8.39), (8.40) give back the results of [52], with the observation that any multiple of $\pi_{2m}^{(4)}(x)$ can be added to $\pi_{2m+1}^{(4)}(x)$. The normalization constant is given by

$$g_{2m}^{(4)} = g_{2m+1}^{(4)} = -h_{2m}^{(a)}. \tag{8.41}$$

The results (8.39), (8.40) derive from (8.37), (8.38) from the indefinite integral,

$$\int L_j^{(a)}(x)\,\mathrm{d}x = -L_{j+1}^{(a-1)}(x) = -L_{j+1}^{(a)}(x) + L_j^{(a)}(x), \tag{8.42}$$

the constants of integration in (8.39), (8.40) being zero on skew-orthogonality with $\pi_1^{(4)}(x)$.

Note. The minus sign in (8.41), where the weight function is $w_a(x)$ disappears in (11.32) where the weight function is $w_a^2(x)$.

To obtain the asymptotic formula for associated Laguerre skew-orthogonal polynomials, we use (8.39), (8.40) and the asymptotic formula for the associated Laguerre orthogonal polynomials (5.6) and (5.7). To avoid θ floating inside the argument, for a given x, we choose θ which effectively corresponds to $j = 2m + \frac{1}{2}$ in (5.6). For a arbitrary and real, and ϵ a fixed positive number, we have for

$$x = (8m + 2a + 4)\cos^2\theta, \qquad \epsilon \le \theta \le \pi/2 - \epsilon m^{-1/2}, \qquad \theta \equiv \theta_{2m+1/2},$$

$$(8.43)\quad \sqrt{g_{2m}^{(4)}}\phi_{2m}^{(4)}(x) = -\frac{(2m)^{a/2}}{2}\left\{\left[\frac{1}{2\sqrt{\pi m \cos\theta \sin^3\theta}}\cos[f_{2m}(\theta)] + 1 + \frac{O(1)}{m\sqrt{x}}\right]\right\},$$

$$(8.44)\quad \sqrt{g_{2m}^{(4)}}\phi_{2m+1}^{(4)}(x) = \frac{(2m)^{a/2}}{\sqrt{m\pi \tan\theta}}\left[\sin[f_{2m}(\theta)] + \frac{O(1)}{\sqrt{mx}}\right],$$

$$(8.45)\quad \sqrt{g_{2m}^{(4)}}\psi_{2m}^{(4)}(x) = (2m)^{a/2}\left[\frac{1}{8\sqrt{\pi m \cos^3\theta \sin\theta}}\sin[f_{2m}(\theta)] + \frac{O(1)}{m\sqrt{x}}\right],$$

$$(8.46)\quad \sqrt{g_{2m}^{(4)}}\psi_{2m+1}^{(4)}(x) = (2m)^{a/2}\left[\frac{1}{2}\sqrt{\frac{\tan\theta}{m\pi}}\cos[f_{2m}(\theta)] + \frac{O(1)}{m\sqrt{x}}\right],$$

with

$$(8.47)\quad \sqrt{g_{2m\pm 2}^{(4)}}\phi_{2m\pm 2}^{(4)}(x) = -\frac{(2m)^{a/2}}{2}\left\{\left[\frac{1}{2\sqrt{\pi m \cos\theta \sin^3\theta}}\cos[f_{2m}(\theta) \mp 4\theta] + 1 + \frac{O(1)}{m\sqrt{x}}\right]\right\}.$$

Similar relations hold for (8.44), (8.45) and (8.46) as $m \to m \pm 1$. Here

$$(8.48)\quad f_{2m}(\theta) = (2m + 1 + a/2)(\sin 2\theta - 2\theta) + \frac{3\pi}{4}.$$

In deriving (8.43) we have used the large-m approximation

$$(8.49)\quad \left(\frac{8m - x}{2m}\right)\pi_{2m}^{(4)}(x) = -\left(L_{2m}^{(a)}(x) + L_{2m+1}^{(a)}(x)\right) - \frac{1}{2}\left(\frac{8m - x}{2m}\right)(2m)^{a/2}\left(w_a(x)\right)^{-1/2},$$

which follows from the three-term recursion and a sum rule for $L_j^{(a)}(x)$ [77].

Level density and 2-point correlation for associated Laguerre symplectic ensemble. In this section, we calculate the level density of the associated Laguerre symplectic ensemble and show that in the bulk of the spectrum, the scaled or "unfolded" [53] kernel function $S_{2N}^{(4)}(x,y)/S_{2N}^{(4)}(x,x)$ is universal.

From (3.55), we need to calculate $R_{2N-1,2N+1}^{(4)}$, $R_{2N-1,2N}^{(4)}$, $R_{2N-2,2N}^{(4)}$ and $P_{2N-1,2N}^{(4)}$. Using the three-term recursion relation in the quaternion space, i.e., (3.39) and (3.40) with $f(x) = x$, we get

$$(8.50)\quad x\frac{\mathrm{d}}{\mathrm{d}x}\phi_{2m+1}^{(4)}(x) = \sum_{j=2m-2}^{2m+2} P_{2m+1,j}^{(4)}\phi_j^{(4)}(x),$$

$$(8.51)\quad \begin{aligned} x^2\frac{\mathrm{d}}{\mathrm{d}x}\phi_{2m+1}^{(4)}(x) &= \sum_{j=2m-2}^{2m+3} R_{2m+1,j}^{(4)}\phi_j^{(4)}(x),\\ x^2\frac{\mathrm{d}}{\mathrm{d}x}\phi_{2m}^{(4)}(x) &= \sum_{j=2m-2}^{2m+2} R_{2m,j}^{(4)}\phi_j^{(4)}(x),\end{aligned}$$

from which, using Table 1, we get

$$P^{(4)}_{2m+1,2m+2} = \sqrt{\frac{g^{(4)}_{2m+2}}{g^{(4)}_{2m}}}(m+1), \tag{8.52}$$

$$R^{(4)}_{2m+1,2m+3} = -\sqrt{\frac{g^{(4)}_{2m+2}}{g^{(4)}_{2m}}}(m+1)(2m+3), \tag{8.53}$$

$$R^{(4)}_{2m,2m+2} = -\sqrt{\frac{g^{(4)}_{2m+2}}{g^{(4)}_{2m}}}(m+1)(2m+1), \tag{8.54}$$

$$R^{(4)}_{2m+1,2m+2} = \sqrt{\frac{g^{(4)}_{2m+2}}{g^{(4)}_{2m}}}(m+1)(4m+a+4). \tag{8.55}$$

For $m = N-1$, large N, we have $g^{(4)}_{2N}/g^{(4)}_{2N-2} \simeq 1$. Finally, for large N, the recursion coefficients are given by

$$\begin{aligned} P^{(4)}_{2N-1,2N} &\sim N + O(1), & R^{(4)}_{2N-1,2N+1} &\sim -2N^2 + O(N), \\ R^{(4)}_{2N-2,2N} &\sim -2N^2 + O(N), & R^{(4)}_{2N-1,2N} &\sim 4N^2 + O(N). \end{aligned} \tag{8.56}$$

Also, since $g^{(4)}_{2N-2}/g^{(4)}_{2N} \simeq 1$, we can write

$$\begin{aligned} \phi^{(4)}_{2N}(x)\phi^{(4)}_{2N-1}(y) &= \frac{1}{\sqrt{g^{(4)}_{2N-2}g^{(4)}_{2N}}}\sqrt{w(x)w(y)}\pi^{(4)}_{2N}(x)\pi^{(4)}_{2N-1}(y) \\ &\simeq \frac{1}{g^{(4)}_{2N}}\sqrt{w(x)w(y)}\pi^{(4)}_{2N}(x)\pi^{(4)}_{2N-1}(y) \\ &= -\frac{1}{h^a_{2N}}\sqrt{w(x)w(y)}\pi^{(4)}_{2N}(x)\pi^{(4)}_{2N-1}(y) \\ &\simeq -(2N)^{-a}\sqrt{w(x)w(y)}\pi^{(4)}_{2N}(x)\pi^{(4)}_{2N-1}(y). \end{aligned} \tag{8.57}$$

Finally, using the asymptotic results (8.43), (8.44), (8.47), defined in the θ interval $[\epsilon, \pi/2 - \epsilon N^{-1/2}]$, and the large m results (8.56) and (8.57) in the generalized Christoffel – Darboux formula (3.55), we get for

$$x = (8N + 2a + 4)\cos^2\theta, \qquad y = x + \Delta x,$$

$$
\begin{aligned}
(8.58)\quad & y(y-x)S_{2N}^{(4)}(x,y) \\
&= R_{2N-2,2N}^{(4)}\left[\phi_{2N}^{(4)}(x)\phi_{2N-1}^{(4)}(y) - \phi_{2N}^{(4)}(y)\phi_{2N-1}^{(4)}(x)\right] \\
&\quad + R_{2N-1,2N+1}^{(4)}\left[\phi_{2N-2}^{(4)}(x)\phi_{2N+1}^{(4)}(y) - \phi_{2N-2}^{(4)}(y)\phi_{2N+1}^{(4)}(x)\right] \\
&\quad + (R_{2N-1,2N}^{(4)} - xP_{2N-1,2N}^{(4)})\left[\phi_{2N-2}^{(4)}(x)\phi_{2N}^{(4)}(y) - \phi_{2N-2}^{(4)}(y)\phi_{2N}^{(4)}(x)\right] \\
&= -\frac{N}{2\pi\sin^2\theta}\{\sin[f_{2N-2}(\theta+\Delta\theta) - f_{2N}(\theta)] \\
&\qquad\qquad + \sin[f_{2N}(\theta+\Delta\theta) - f_{2N-2}(\theta)]\} \\
&\quad + \frac{N\cos 2\theta}{4\pi\cos\theta\sin^3\theta}\{\cos[f_{2N-2}(\theta)]\cos[(f_{2N}(\theta+\Delta\theta)] \\
&\qquad\qquad - \cos[f_{2N-2}(\theta+\Delta\theta)]\cos[f_{2N}(\theta)]]\} \\
&= -\left[\frac{N\cos 4\theta}{\pi\sin^2\theta} - \frac{N\sin 4\theta\cos 2\theta}{4\pi\cos\theta\sin^3\theta}\right]\sin\left(\frac{\partial f_{2N}}{\partial\theta}\Delta\theta\right) \\
&= -N\frac{[\cos 4\theta - \cos^2 2\theta]}{\pi\sin^2\theta}\sin\left(\frac{\partial f_{2N}}{\partial\theta}\Delta\theta\right),
\end{aligned}
$$

which gives

$$\Delta x S_{2N}^{(4)}(x,y) = -\frac{\sin(8N\sin^2\theta\Delta\theta)}{2\pi} = \frac{\sin\left(\frac{1}{2}x^{-1/2}\sqrt{8N-x}\Delta x\right)}{2\pi}. \tag{8.59}$$

Thus we get

$$S_{2N}^{(4)}(x,y) = \frac{\sin\left(\frac{1}{2}x^{-1/2}\sqrt{8N-x}\Delta x\right)}{2\pi\Delta x}, \qquad 0 < x < 8N. \tag{8.60}$$

With $\Delta x \to 0$, we get the level density

$$S_{2N}^{(4)}(x,x) = \frac{1}{4\pi}\sqrt{\frac{8N-x}{x}}, \qquad 0 < x < 8N, \tag{8.61}$$

with

$$\int_0^\infty S_{2N}^{(4)}(x,x)\,\mathrm{d}x = N. \tag{8.62}$$

Finally, the scaled kernel function

$$\frac{S_{2N}^{(4)}(x,y)}{S_{2N}^{(4)}(x,x)} = \frac{\sin 2\pi r}{2\pi r}, \qquad r := \Delta x S_{2N}^{(4)}(x,x), \tag{8.63}$$

is independent of $w_a(x)$ and is the universal sine-kernel.

2.3. Gaussian weight: $\beta = 4$. We finally turn to the Gaussian weight. We have now, with $w(x) = \exp(-2x^2)$,

$$\pi_{2m+1}^{(4)\prime}(x) = 2(2m+1)H_{2m}(x\sqrt{2}), \tag{8.64}$$

$$\pi_{2m}^{(4)\prime}(x) = 4mH_{2m-1}(x\sqrt{2}) + 4m\pi_{2m-2}^{(4)\prime}(x), \tag{8.65}$$

$$\pi_{2m+1}^{(4)}(x) = \frac{1}{\sqrt{2}}H_{2m+1}(x\sqrt{2}), \tag{8.66}$$

$$\pi_{2m}^{(4)}(x) = \frac{1}{\sqrt{2}}H_{2m}(x\sqrt{2}) + 4m\pi_{2m-2}^{(4)}(x), \tag{8.67}$$

along with the normalization

$$g_{2m}^{(4)} = g_{2m+1}^{(4)} = (2m+1)!\pi^{1/2}2^{2m}. \tag{8.68}$$

For large m, we have, with $x = (2m+3/2)^{1/2}\cos\theta$ corresponding to $j = 2m+1$ and $x \to x\sqrt{2}$ in (5.11),

$$\phi_{2m+1}^{(4)}(x) = \frac{1}{m^{1/4}\sqrt{\pi\sin\theta}}\sin\left[\left(m+\frac{3}{4}\right)(\sin 2\theta - 2\theta) + \frac{3\pi}{4}\right], \tag{8.69}$$

$$\phi_{2m}^{(4)}(x) = \frac{1}{(4m)^{1/4}}\left[\frac{1}{2\sqrt{2m\pi\sin^3\theta}}\cos\left[\left(m+\frac{3}{4}\right)(\sin 2\theta - 2\theta) + \frac{3\pi}{4}\right] + \frac{1}{2}\right], \tag{8.70}$$

$$\phi_{2m+1}^{(4)}{}'(x) = 2m^{1/4}\sqrt{\frac{2\sin\theta}{\pi}}\cos\left[\left(m+\frac{3}{4}\right)(\sin 2\theta - 2\theta) + \frac{3\pi}{4}\right], \tag{8.71}$$

$$\phi_{2m}^{(4)}{}'(x) = -\frac{1}{(m)^{1/4}\sqrt{2\pi\sin\theta}}\sin\left[\left(m+\frac{3}{4}\right)(\sin 2\theta - 2\theta) + \frac{3\pi}{4}\right], \tag{8.72}$$

with

$$\begin{aligned}\phi_{2m\pm 2}^{(4)}(x) \\ = \frac{1}{(4m)^{1/4}}\left[\frac{1}{2\sqrt{2m\pi\sin^3\theta}}\cos\left[\left(m+\frac{3}{4}\right)(\sin 2\theta - 2\theta) \mp \theta + \frac{3\pi}{4}\right] + \frac{1}{2}\right].\end{aligned} \tag{8.73}$$

Similar relations hold for (8.69), (8.71), (8.72) as $m \to m \pm 1$.

Here, in deriving the asymptotic behavior of the even polynomial, we have used the large m approximation,

$$\begin{aligned}2\sqrt{2}\pi_{2m}^{(4)}(x) \\ = \frac{1}{4(2m-x^2)}\left[-H_{2m+2}(x\sqrt{2}) + 4mH_{2m}(x\sqrt{2})\right] + m^{-1/4}\left(g_{2m}^{(4)}\right)^{1/2}\mathrm{e}^{x^2}.\end{aligned} \tag{8.74}$$

Level density and 2-point correlation for GSE. In this section, we calculate the level density of the Gaussian symplectic ensemble and show that in the bulk of the spectrum, the scaled or "unfolded" [53] kernel function $S_{2N}^{(4)}(x,y)/S_{2N}^{(4)}(x,x)$ is universal.

From (3.55), we need to calculate $R_{2N-1,2N+1}^{(4)}$, $R_{2N-1,2N}^{(4)}$, $R_{2N-2,2N}^{(4)}$ and $P_{2N-1,2N}^{(4)}$. We have the recursion relations

$$\frac{\mathrm{d}}{\mathrm{d}x}\phi_{2m+1}^{(4)}(x) = \sum_{j=2m-2}^{2m+2} P_{2m+1,j}^{(4)}\phi_j^{(4)}(x), \tag{8.75}$$

$$\begin{aligned} x\frac{\mathrm{d}}{\mathrm{d}x}\phi_{2m+1}^{(4)}(x) &= \sum_{j=2m-2}^{2m+3} R_{2m+1,j}^{(4)}\phi_j^{(4)}(x), \\ x\frac{\mathrm{d}}{\mathrm{d}x}\phi_{2m}^{(4)}(x) &= \sum_{j=2m-2}^{2m+2} R_{2m,j}^{(4)}\phi_j^{(4)}(x). \end{aligned} \tag{8.76}$$

Using (8.75) and (8.76), we get

$$P^{(4)}_{2m+1,2m+2} = -\sqrt{\frac{g^{(4)}_{2m+2}}{2g^{(4)}_{2m}}}, \tag{8.77}$$

$$R^{(4)}_{2m+1,2m+3} = -\frac{1}{4}\sqrt{\frac{g^{(4)}_{2m+2}}{g^{(4)}_{2m}}}, \tag{8.78}$$

$$R^{(4)}_{2m,2m+2} = -\frac{1}{4}\sqrt{\frac{g^{(4)}_{2m+2}}{g^{(4)}_{2m}}}, \tag{8.79}$$

$$R^{(4)}_{2m+1,2m+2} - 0. \tag{8.80}$$

For $m = N-1$, large N, we have $\left(g^{(4)}_{2N}/g^{(4)}_{2N-2}\right)^{1/2} \simeq 4N$. Hence for large N, the recursion coefficients are given by

$$\begin{aligned} &P^{(4)}_{2N-1,2N} \sim -2\sqrt{2}N + O(1), && R^{(4)}_{2N-1,2N+1} \sim -N + O(1), \\ &R^{(4)}_{2N-2,2N} \sim -N + O(1), && R^{(4)}_{2N-1,2N} = 0. \end{aligned} \tag{8.81}$$

Finally, using the asymptotic results (8.69) – (8.72), defined in the θ interval $[\epsilon, \pi - \epsilon]$, and the large m results (8.81) in the generalized Christoffel – Darboux formula (3.55), we get for

$$x = (2N + \tfrac{3}{2})^{1/2}\cos\theta, \qquad y = x + \Delta x,$$

$$\begin{aligned} (y-x)S^{(4)}_{2N}(x,y) &= R^{(4)}_{2N-2,2N}\left[\phi^{(4)}_{2N}(x)\phi^{(4)}_{2N-1}(y) - \phi^{(4)}_{2N}(y)\phi^{(4)}_{2N-1}(x)\right] \\ &\quad + R^{(4)}_{2N-1,2N+1}\left[\phi^{(4)}_{2N-2}(x)\phi^{(4)}_{2N+1}(y) - \phi^{(4)}_{2N-2}(y)\phi^{(4)}_{2N+1}(x)\right] \\ &\quad - xP^{(4)}_{2N-1,2N}\left[\phi^{(4)}_{2N-2}(x)\phi^{(4)}_{2N}(y) - \phi^{(4)}_{2N-2}(y)\phi^{(4)}_{2N}(x)\right] \\ &= -\frac{1}{4\pi\sin^2\theta}\{\sin[f_{2N-2}(\theta+\Delta\theta) - f_{2N}(\theta)] \\ &\qquad\qquad + \sin[f_{2N}(\theta+\Delta\theta) - f_{2N-2}(\theta)]\} \\ &\quad + \frac{\cos\theta}{4\pi\sin^3\theta}\{\cos[f_{2N-2}(\theta)]\cos[f_{2N}(\theta+\Delta\theta)] \\ &\qquad\qquad - \cos[f_{2N-2}(\theta+\Delta\theta)]\cos[f_{2N}(\theta)]\} \\ &= -\frac{1}{2\pi\sin^2\theta}[\cos 2\theta - \cos^2\theta]\sin\left(\frac{\partial f_{2N}(\theta)}{\partial\theta}\Delta\theta\right) \\ &= \frac{1}{2\pi}\sin\left(\frac{\partial f_{2N}(\theta)}{\partial\theta}\Delta\theta\right), \end{aligned} \tag{8.82}$$

where

$$f_{2m}(\theta) = \left(m + \frac{3}{4}\right)(\sin 2\theta - 2\theta) + \frac{3\pi}{4}. \tag{8.83}$$

We have

$$\Delta x S^{(4)}_{2N}(x,y) = \frac{\sin(2\sqrt{2N - x^2}\Delta x)}{2\pi}, \tag{8.84}$$

such that

$$S_{2N}^{(4)}(x,y) = \frac{\sin\left(2\sqrt{2N-x^2}\Delta x\right)}{2\pi\Delta x}, \qquad |x| < \sqrt{2N}. \tag{8.85}$$

With $\Delta x \to 0$, we get the level density

$$S_{2N}^{(4)}(x,x) = \frac{1}{\pi}\sqrt{2N-x^2}, \qquad |x| < \sqrt{2N}, \tag{8.86}$$

with

$$\int_{-\infty}^{\infty} S_{2N}^{(4)}(x,x)\, \mathrm{d}x = N. \tag{8.87}$$

Finally, the scaled kernel function

$$\frac{S_{2N}^{(4)}(x,y)}{S_{2N}^{(4)}(x,x)} = \frac{\sin 2\pi r}{2\pi r}, \qquad r := \Delta x S_{2N}^{(4)}(x,x), \tag{8.88}$$

is independent of the weight function and is the universal sine-kernel.

3. Quartic two-cut weight: $\beta = 4$

In this section, we use the definition (2.35) to study the asymptotic properties of skew-orthogonal functions arising in the study of symplectic ensembles corresponding to the quartic two-cut potential.

With $m, N \to \infty$ and in the range $x_1 + \delta < x < x_2 - \delta$, $\delta > 0$, and neglecting the m-independent term, we get using the mapping relations (4.79) and the asymptotic results for the corresponding orthogonal functions (5.16):

$$\begin{aligned}
\phi_{2m+1}^{(4)}(x) &\simeq 2C\sqrt{\frac{L}{x\sin\theta}}[\cos[f_{2m+2}(\theta)] + O(1/N)],\\
\phi_{2m}^{(4)}(x) &\simeq -\frac{C}{N\lambda'}\sqrt{\frac{Rx}{\sin^3\theta}}[\sin[f_{2m+2}(\theta)] + O(1/N)],\\
\psi_{2m+1}^{(4)}(x) &\simeq -\left[\frac{4NC\lambda'}{\sqrt{R}}\right]\sqrt{x\sin\theta}[\sin[f_{2m+2}(\theta)] + O(1/N)],\\
\psi_{2m}^{(4)}(x) &\simeq -2C\sqrt{\frac{x^3}{L\sin\theta}}[\cos[f_{2m+2}(\theta)] + O(1/N)],
\end{aligned} \tag{8.89}$$

where $f_m(\theta)$, $C_m \equiv C$, χ, λ and θ are defined in (5.16) and (6.63). Here, we note, that for a given x, θ varies with m. We have chosen $\theta \equiv \theta_{2m+2}$ such that

$$\phi_{2m+1\pm k}^{(4)}(x) \simeq 2C\sqrt{\frac{L}{x\sin\theta}}\left[\cos\left[f_{2m+2}(\theta) \mp \frac{k\theta}{2}\right] + O(1/N)\right], \qquad m \gg k, \tag{8.90}$$

and so on for $\phi_{2m\pm k}^{(4)}(x)$, $\psi_{2m+1\pm k}^{(4)}(x)$, $\psi_{2m\pm k}^{(4)}(x)$.

For the recursion coefficients $P_{j,k}^{(4)}$ and $R_{j,k}^{(4)}$ (as defined in Chapter 3), we simply read them off from (4.74)–(4.77) using the two-term recursion relation for the corresponding orthogonal polynomials (4.4). For large m we have

$$\begin{aligned} P^{(4)}_{2m,2m+3} &= -\sqrt{\frac{R}{L}}, & P^{(4)}_{2m+1,2m+4} &= -a^2 L,\\ P^{(4)}_{2m,2m+1} &= \frac{t}{gL}, & P^{(4)}_{2m+1,2m+2} &= 0,\\ R^{(4)}_{2m,2m+4} &= R^{(4)}_{2m+1,2m+5} = -N\lambda, \\ R^{(4)}_{2m,2m+2} &= R^{(4)}_{2m+1,2m+3} = Nt\sqrt{RL}, \end{aligned} \tag{8.91}$$

where $a = -N\sqrt{g\lambda}$.

To calculate $S^{(4)}_{2N}(x,y)$, we use (8.89), (8.90) and (8.91) in (3.59) for $\beta = 4$. Here $f_{2N+2}(\theta) \equiv f(\theta)$, the kernel function is given by (3.59)(also see [36]). The first four terms give

$$\begin{aligned} \frac{2C^2_{2N+2}\sqrt{LR}}{\sin^2\theta}\big[&\cos(f(\theta)+2\theta)\sin\big(f(\theta+\Delta\theta)\big) - \sin\big(f(\theta)\big)\cos(f(\theta+\Delta\theta)+2\theta)\\ &+\cos(f(\theta)+\theta)\sin(f(\theta+\Delta\theta)-\theta) - \cos(f(\theta+\Delta\theta)+\theta)\sin\big(f(\theta)-\theta\big)\\ &+\sin(f(\theta+\Delta\theta)+2\theta)\cos\big(f(\theta)\big) - \sin(f(\theta)+2\theta)\cos\big(f(\theta+\Delta\theta)\big)\\ &+\sin(f(\theta+\Delta\theta)+\theta)\cos(f(\theta)-\theta) - \sin(f(\theta)+\theta)\cos(f(\theta+\Delta\theta)-\theta)\big]\\ &= \frac{2\cos 2\theta\sin\big(\Delta\theta f'(\theta)\big)}{\pi\sin^2\theta} \end{aligned}$$

Following similar algebra, the next four terms give

$$\begin{aligned} -\frac{4C^2_{2N+2}t\cos\theta\sin(\Delta\theta f'(\theta))}{g\sin^2\theta} - \frac{4y^2gC^2_{2N+2}\cos\theta\sin(\Delta\theta f'(\theta))}{g\sin^2\theta}&\\ = -\frac{4C^2_{2N+2}\cos\theta\sin(\Delta\theta f'(\theta))}{g\sin^2\theta}\left[\frac{gy^2+t}{2\sqrt{\lambda' g}}\right]2\sqrt{\lambda' g}&\\ = -\frac{2\cos^2\theta\sin(\Delta\theta f'(\theta))}{\pi\sin^2\theta}& \end{aligned} \tag{8.92}$$

while the last term gives $\sin(\Delta\theta f'(\theta))/\pi$. Here, we have used the asymptotic results (as $m, N \to \infty$):

$$\begin{aligned} q = \cos\theta = \frac{gy^2+t}{2\sqrt{\lambda' g}}, \qquad C_m = \frac{1}{2\sqrt{\pi}}\left(\frac{g}{\lambda}\right)^{1/4}(1+O(N^{-1})),\\ \lambda' = \lambda + O(N^{-1}). \end{aligned} \tag{8.93}$$

Combining all, we get for $x = y + \Delta y$,

$$\begin{aligned} (y-x)S^{(4)}_{2N}(x,y) &= -\frac{\sin(\Delta\theta f'(\theta))}{\pi} + O(N^{-1})\\ \Longrightarrow S^{(4)}_{2N}(x,y) &= \frac{\sin\big[\Delta y 2N\sqrt{g}y\sqrt{1-q^2}\big]}{\pi\Delta y} + O(N^{-1}). \end{aligned} \tag{8.94}$$

For level-density, we take the limit

$$\lim_{\Delta y\to 0}\frac{1}{2N}S_{2N}^{(4)}(x,y)=\frac{1}{2N}S_{2N}^{(4)}(y,y)=\frac{\sqrt{g}}{\pi}|y|\sqrt{1-q^2} \tag{8.95}$$
$$=\frac{|y|}{\pi}\sqrt{g-\left(\frac{gy^2+t}{2}\right)^2}+O(N^{-1}),$$

with

$$\int_{-\infty}^{\infty}\frac{S_{2N}^{(4)}(y,y)}{2}\,dy=N, \tag{8.96}$$

such that we get the "universal" sine-kernel

$$\frac{S_{2N}^{(4)}(x,y)}{S_{2N}^{(4)}(y,y)}=\frac{\sin 2\pi r}{2\pi r},\qquad r=\Delta y S_{2N}^{(4)}(y,y)/2. \tag{8.97}$$

Exercise. Readers who do not like the "2" floating in the density are advised to repeat the above proof of universality using skew-orthogonal functions satisfying the normalization $\left(\phi_j^{(4)}(x),\psi_k^{(4)}(x)\right)=Z_{j,k}/2$, instead of $Z_{j,k}$, (i.e., definition (2.34) instead of (2.35)) keeping in mind the concept of "unfolding" of the spectra. (Hint: See [39].)

Summary. In Chapters 6 – 8 we have used the mapping relations obtained in Chapter 4 to derive the bulk asymptotics of skew-orthogonal functions and their recursion coefficients $P_{j,k}^{(\beta)}$ and $R_{j,k}^{(\beta)}$ for $\beta=1$ and 4. As an application in random matrix theory, we use these asymptotic results in the generalized Christoffel – Darboux formula, obtained in Chapter 3, to derive the level densities and prove universality of the correlation functions.

CHAPTER 9

Skew-orthogonal Polynomials and Differential Systems

In this chapter, we show that a finite subsequence of skew-orthonormal vectors $\Phi_W^{(\beta)}(x)$ and $\Psi_W^{(\beta)}(x)$, defined with respect to the weight function $w(x) = \exp\bigl(-2V(x)\bigr)$, satisfy a mutually compatible set of difference-differential-deformation equations. $V(x)$, as defined in (3.25), is a polynomial of degree $2d$ with positive leading coefficient u_{2d}. We call the coefficients u_K the deformation parameters. The 'size' or rank of this finite subsequence is determined by the generalized Christoffel – Darboux formula which gives an indication of the number of terms near the so-called 'Fermi-level' which are actually required to study this system. We also find that the mutually compatible system of differential-difference-deformation equation satisfied by the finite subsequence of skew-orthogonal vectors admits a common fundamental solution.

In this context, we may recall that, in Chapters 3 and 4, we have not only derived the generalized Christoffel – Darboux formula for skew-orthogonal functions but also mapped them to their more illustrious orthogonal functions. This helped in deriving the asymptotic properties of skew-orthogonal functions. The natural question that may be asked is "if you have to ultimately revert to orthogonal polynomials to find the asymptotics of skew-orthogonal polynomials, then why not use the Sener – Verbaarschot – Widom formalism [72, 78] itself!" Up to this date the answer to this pertinent question is that skew-orthogonal functions trigger a (i) drastic simplification of random matrix calculations and (ii) help in our understanding of some of the finer aspects of orthogonal and symplectic ensembles.

In this chapter, we also address the issue of the existence of the fundamental solution of a system of differential-difference-deformation equation. A Riemann – Hilbert analysis, using the method of steepest descent (as done for orthogonal polynomials: see [18] and references therein) would be the next step to derive (and hence verify results obtained using mapping relations) the asymptotics of skew-orthogonal functions.

One of the by-products is that we observe a certain duality between skew-orthonormal vectors corresponding to $\beta = 1$ and 4. This will be discussed in Chapter 11.

The existence of difference equations (9.35), (9.36), deformation equations (9.81) and differential equations (9.89) for vectors arising in the orthogonal and symplectic ensembles of random matrices can be viewed as just a projection of the semi-infinite functions $\Phi^{(\beta)}(x)$ and $\Psi^{(\beta)}(x)$ onto the finite "window" $\Phi_W^{(\beta)}(x)$ and $\Psi_W^{(\beta)}(x)$. However, we may also consider these equations as defining an overdetermined system of finite difference-differential-deformation equations of the vector functions and see that these systems are compatible. This leads to the existence of a fundamental matrix solution, denoted by $\Phi^{(\beta)}_{[W]}(x)$ and $\Psi^{(\beta)}_{[W]}(x)$, where all the column vectors are simultaneous solutions of the system of difference-differential-deformation equations.

Compatibility of the deformation and difference equation with the differential equation implies that the generalized monodromy of the differential operators are invariant under u_K deformations and shifts in N.

1. Recursion relations and finite band matrices

In this section, we revisit the topic of recursion relations satisfied by the skew-orthonormal vectors with polynomial weight. Unlike orthogonal functions, skew-orthonormal vectors do not satisfy a three-term recursion relation and hence do not give birth to the tridiagonal Jacobi matrix. Instead, they give rise to a $(2d+1)$-term recursion relation, where $2d$ is the degree of the polynomial $V(x)$ (see (3.25)). We have

$$x\Phi^{(\beta)}(x) = Q^{(\beta)}\Phi^{(\beta)}(x), \tag{9.1}$$

$$\frac{\mathrm{d}}{\mathrm{d}x}\Phi^{(\beta)}(x) = P^{(\beta)}\Phi^{(\beta)}(x), \qquad \frac{\mathrm{d}}{\mathrm{d}x}\Psi^{(\beta)}(x) = P^{(\beta)}\Psi^{(\beta)}(x), \qquad \beta = 1,4, \tag{9.2}$$

$$x\frac{\mathrm{d}}{\mathrm{d}x}\Phi^{(4)}(x) = R^{(4)}\Phi^{(4)}(x), \qquad x\frac{\mathrm{d}}{\mathrm{d}x}\Psi^{(4)}(x) = \left[R^{(4)} - \mathbf{1}\right]\Psi^{(4)}(x), \tag{9.3}$$

$$x\frac{\mathrm{d}}{\mathrm{d}x}\Psi^{(1)}(x) = R^{(1)}\Psi^{(1)}(x), \qquad x\frac{\mathrm{d}}{\mathrm{d}x}\Phi^{(1)}(x) = \left[R^{(1)} - \mathbf{1}\right]\Phi^{(1)}(x), \tag{9.4}$$

where $\Phi^{(\beta)}(x)$ and $\Psi^{(\beta)}(x)$ are the semi-infinite column vectors defined in (3.16), (3.17). This is equivalent to saying

$$[R^{(4)} - xP^{(4)}]\Phi^{(4)}(x) = 0, \qquad [R^{(4)} - xP^{(4)} - \mathbf{1}]\Psi^{(4)}(x) = 0, \tag{9.5}$$

$$[R^{(1)} - xP^{(1)}]\Psi^{(1)}(x) = 0, \qquad [R^{(1)} - xP^{(1)} - \mathbf{1}]\Phi^{(1)}(x) = 0. \tag{9.6}$$

These semi-infinite matrices satisfy the following relations

$$R^{(4)} = P^{(4)}Q^{(4)}, \qquad R^{(1)} = Q^{(1)}P^{(1)}, \tag{9.7}$$

and the commutation relations:

$$[Q^{(\beta)}, P^{(\beta)}] = \mathbf{1}, \qquad [R^{(\beta)}, P^{(\beta)}] = P^{(\beta)}. \tag{9.8}$$

Here, each entry is a 2×2 quaternion of the form

$$A_{n,m} := \begin{bmatrix} \tilde{A}_{2n,2m} & \tilde{A}_{2n,2m+1} \\ \tilde{A}_{2n+1,2m} & \tilde{A}_{2n+1,2m+1} \end{bmatrix}. \tag{9.9}$$

Using $(\psi_n^{(4)}, \psi_m^{(4)})$ and $(x\psi_n^{(4)}, \psi_m^{(4)})$ for $\beta = 4$, and replacing $\psi^{(4)}(x)$ by $\phi^{(1)}(x)$ for $\beta = 1$, we get the $2d+1$ band quaternion matrices which satisfy the anti-self dual condition:

$$P^{(\beta)} = -P^{(\beta)\mathrm{D}}, \quad R^{(\beta)} = -R^{(\beta)\mathrm{D}}, \qquad \beta = 1,4, \tag{9.10}$$

where the dual of a matrix A is defined in (2.6).

Remark. Anti-self-duality condition imposes a much weaker restriction on a diagonal quaternion than the antisymmetry condition on a diagonal matrix element. For example, it leaves the off-diagonal entry of the diagonal quaternion arbitrary. This is the reason why the odd skew-orthogonal polynomials are arbitrary up to the addition of a lower even order polynomial.

However, starting with $(x\phi_n^{(\beta)}, \psi_m^{(\beta)})$ and using (9.8), we get

$$(Q^{(\beta)} - Q^{(\beta)\mathrm{D}})P^{(\beta)} = P^{(\beta)}(Q^{(\beta)} - Q^{(\beta)\mathrm{D}}) = \mathbf{1}. \tag{9.11}$$

As pointed out in the beginning of the section, this essentially means that like orthogonal polynomials, we do not have a tridiagonal Jacobi matrix for skew-orthogonal polynomials. It is this relation that causes the generalized Christoffel–Darboux sum to have a local behavior (i.e., the rank has dependence on the measure or weight function). In a sense, this is a major setback to our hope of defining a 2×2 matrix Riemann–Hilbert problem for skew-orthogonal vectors, similar to that of orthogonal polynomials. Also, from the dependence of $P^{(\beta)}$ and $R^{(\beta)}$ on $2d$ (the degree of the polynomial $V(x)$), we can conclude that the 'size' of our q-matrix Riemann–Hilbert problem depends on d.

We also have [36] for $\beta = 4$:

$$\begin{aligned} \phi_n'^{(4)}(x) &= \sum P_{n,m}^{(4)}\phi_m^{(4)}(x) \\ &= \left[-V'(x)\phi_n^{(4)}(x) + \phi_{n-1}^{(4)}(x) + \cdots\right] \\ &= -\sum\left(V'(Q^{(4)})\right)_{n,m}\phi_m^{(4)}(x) + \phi_{n-1}^{(4)}(x) + \cdots, \end{aligned} \tag{9.12}$$

while for $\beta = 1$, we have

$$\begin{aligned} \phi_n^{(1)}(x) &= \sum P_{n,m}^{(1)}\psi_m^{(1)}(x) \\ &= \int \frac{\mathrm{d}}{\mathrm{d}y}\left[\phi_n^{(1)}(y)\right]\epsilon(x-y)\ \mathrm{d}y \\ &= \int\left[-\sum(V'(Q^{(1)}))_{n,m}\phi_m^{(1)}(y)\right]\epsilon(x-y)\ \mathrm{d}y + \psi_{n-1}^{(1)}(x) + \cdots. \end{aligned} \tag{9.13}$$

This gives

$$P^{(\beta)} + [V'(Q^{(\beta)})] = \text{lower}. \tag{9.14}$$

Similarly, for $\beta = 4$, we have

$$\begin{aligned} x\phi_n'^{(4)}(x) &= \sum R_{n,m}^{(4)}\phi_m^{(4)}(x) \\ &= \left[-xV'(x)\phi_n^{(4)}(x) + n\phi_n^{(4)}(x) + \cdots\right] \\ &= -\sum_{m,l} Q_{n,m}^{(4)}\left(V''(Q^{(4)})\right)_{m,l}\phi_l^{(4)}(x) + \cdots, \end{aligned} \tag{9.15}$$

while for $\beta = 1$, we have

$$\begin{aligned} x\phi_n^{(1)}(x) &= \sum R_{n,m}^{(1)}\psi_m^{(1)}(x) \\ &= \int \frac{\mathrm{d}}{\mathrm{d}y}\left[y\phi_n^{(1)}(y)\right]\epsilon(x-y)\ \mathrm{d}y \\ &= \left[-\sum_{m,l} Q_{n,m}^{(1)}\left(V'(Q^{(1)})\right)_{m,l}\psi_l^{(1)}(x)\right] + \cdots, \end{aligned}$$

which gives us

$$R^{(\beta)} + [Q^{(\beta)} V'(Q^{(\beta)})] = \text{lower}_+, \tag{9.16}$$

where 'lower' and 'lower$_+$' denote a strictly lower triangular matrix and a lower triangular matrix with the principal diagonal respectively. This, along with the anti-self duality condition ensures that the matrices $P^{(\beta)}$ and $R^{(\beta)}$ have finite bands above and below the principal diagonal.

Having obtained the recursion relations for skew-orthonormal vectors, we introduce a convenient notation to express the band matrices $P^{(\beta)}$ and $R^{(\beta)}$:

$$\begin{aligned} \sum_{m=n-d}^{n+d} P^{(\beta)}_{nm}\Phi^{(\beta)}_m(x) &= \sum_{k=-d}^{d} \zeta^{(\beta)}_k(n)\Phi^{(\beta)}_{n-k}(x), \\ \sum_{m=n-d}^{n+d} R^{(\beta)}_{nm}\Phi^{(\beta)}_m(x) &= \sum_{k=-d}^{d} \eta^{(\beta)}_k(n)\Phi^{(\beta)}_{n-k}(x), \end{aligned} \tag{9.17}$$

and

$$\sum_{m=n-d}^{n+d} R^{(\beta)}_{nm}\Phi^{(\beta)}_m(x) - x\sum_{m=n-d}^{n+d} P^{(\beta)}_{nm}\Phi^{(\beta)}_m(x) = \sum_{k=-d}^{d} \alpha^{(\beta)}_k(n)\Phi^{(\beta)}_{n-k}(x), \tag{9.18}$$

where

$$\begin{aligned} \eta^{(\beta)}_k &:= \operatorname{diag}(\eta^{(\beta)}_k(0),\ldots,\eta^{(\beta)}_k(n),\ldots), \\ \zeta^{(\beta)}_k &:= \operatorname{diag}(\zeta^{(\beta)}_k(0),\ldots,\zeta^{(\beta)}_k(n),\ldots), \\ \alpha^{(\beta)}_k &:= \operatorname{diag}(\alpha^{(\beta)}_k(0),\ldots,\alpha^{(\beta)}_k(n),\ldots). \end{aligned} \tag{9.19}$$

Here, we have suppressed the x-dependence of $\alpha^{(\beta)}_k(n)$. This notation is useful in deriving the folding function needed to project a given sub-sequence of skew-orthogonal vectors onto the finite window.

More explicitly, with this notation, we can write

$$P^{(\beta)} \tag{9.20}$$

$$= \left[\begin{matrix} \zeta^{(\beta)}_0(0) & \zeta^{(\beta)}_{-1}(0) & \cdots & \cdots & \zeta^{(\beta)}_{-d}(0) & 0 & 0 & \cdots \\ \zeta^{(\beta)}_1(1) & \zeta^{(\beta)}_0(1) & \zeta^{(\beta)}_{-1}(1) & \cdots & \zeta^{(\beta)}_{-d+1}(1) & \zeta^{(\beta)}_{-d}(1) & 0 & \cdots \\ \zeta^{(\beta)}_2(2) & \zeta^{(\beta)}_1(2) & \zeta^{(\beta)}_0(2) & \cdots & \zeta^{(\beta)}_{-d+2}(2) & \zeta^{(\beta)}_{-d+1}(2) & \zeta^{(\beta)}_{-d}(2) & 0 \\ \vdots & \vdots & \vdots & \cdots & \cdots & \cdots & \cdots & \ddots \\ \zeta^{(\beta)}_d(d) & \zeta^{(\beta)}_{d-1}(d) & \cdots & \cdots & \zeta^{(\beta)}_0(d) & \cdots & \cdots & \cdots \\ 0 & \zeta^{(\beta)}_d(d+1) & \ddots & \ddots & \ddots & \ddots & \ddots & \cdots \\ 0 & 0 & \zeta^{(\beta)}_d(d+2) & \ddots & \ddots & \ddots & & \end{matrix}\right)$$

and

$$(9.21)\quad R^{(\beta)} = \begin{bmatrix} \eta_0^{(\beta)}(0) & \eta_{-1}^{(\beta)}(0) & \cdots & \cdots & \eta_{-d}^{(\beta)}(0) & 0 & 0 & \cdots \\ \eta_1^{(\beta)}(1) & \eta_0^{(\beta)}(1) & \eta_{-1}^{(\beta)}(1) & \cdots & \eta_{-d}^{(\beta)}(1) & \eta_{-d}^{(\beta)}(1) & 0 & \cdots \\ \eta_2^{(\beta)}(2) & \eta_1^{(\beta)}(2) & \eta_0^{(\beta)}(2) & \cdots & \eta_{-d+1}^{(\beta)}(2) & \eta_{-d}^{(\beta)}(2) & \eta_{-d}^{(\beta)}(2) & 0 \\ \vdots & \vdots & \vdots & \cdots & \cdots & \cdots & \cdots & \ddots \\ \eta_d^{(\beta)}(d) & \eta_{d-1}^{(\beta)}(d) & \cdots & \cdots & \eta_0^{(\beta)}(d) & \cdots & \cdots & \cdots \\ 0 & \eta_d^{(\beta)}(d+1) & \ddots & \ddots & \ddots & \ddots & \ddots & \cdots \\ 0 & 0 & \eta_d^{(\beta)}(d+2) & \ddots & \ddots & \ddots & & \end{bmatrix}.$$

We note that for even $V(x)$, the quaternions of the outermost band are such that

$$(9.22)\ \forall n,\quad \zeta_{\pm d}^{(\beta)}(n) \equiv \begin{bmatrix} 0 & 0 \\ \left(\zeta_{\pm d}^{(\beta)}(n)\right)_{10} & 0 \end{bmatrix}, \quad \eta_{\pm d}^{(\beta)}(n) \equiv \begin{bmatrix} \left(\eta_{\pm d}^{(\beta)}(n)\right)_{00} & 0 \\ \left(\eta_{\pm d}^{(\beta)}(n)\right)_{10} & \left(\eta_{\pm d}^{(\beta)}(n)\right)_{11} \end{bmatrix}.$$

Exercise. Obtain the $d = 1$ result and verify with the recursion relations for Gaussian weight, given in (3.40).

From the commutation relations (9.8), we get the following quadratic relation between the coefficients: $(\zeta_j^{(\beta)}, \eta_j^{(\beta)})$

$$(9.23)\quad \boxed{\sum_{j=l}^{d-1} \eta_{l-j}^{(\beta)}(n)\zeta_j^{(\beta)}(n-l+j) - \sum_{j=l}^{d-1} \zeta_{l-j}^{(\beta)}(n)\eta_j^{(\beta)}(n-l+j) = \zeta_l^{(\beta)}(n).}$$

This is the compatibility relation for the difference-differential system and will be used later.

Remark. Here, one can notice the basic difference between the properties of bi-orthogonal polynomials and their skew-orthogonal counterparts. For skew-orthogonal vectors, the semi-infinite quaternion matrices are symmetric around the principal diagonal. This is not true for bi-orthogonal polynomials where the matrices have an asymmetry [6, 8, 9, 53].

2. The differential system

The deformation matrix. In this section, we consider infinitesimal deformation corresponding to changes in u_K, the coefficient of the potential $V(x)$. Using the definition of the semi-infinite matrices $\Phi^{(\beta)}(x)$ and $\widehat{\Phi}^{(\beta)}(x)$ (and $\Psi^{(\beta)}(x)$, $\widehat{\Psi}^{(\beta)}(x)$), the deformation matrix can be defined as

$$(9.24)\quad \frac{\partial}{\partial u_K}\Phi^{(\beta)}(x) = U_K^{(\beta)}\Phi^{(\beta)}(x), \qquad \frac{\partial}{\partial u_K}\widehat{\Phi}^{(\beta)}(x) = \widehat{\Phi}(x)\left[U_K^{(\beta)}\right]^{\mathrm{D}}, \qquad K = 1, \ldots, 2d,$$

$$(9.25)\quad \frac{\partial}{\partial u_K}\Psi^{(\beta)}(x) = U_K^{(\beta)}\Psi^{(\beta)}(x), \qquad \frac{\partial}{\partial u_K}\widehat{\Psi}^{(\beta)}(x) = \widehat{\Psi}(x)\left[U_K^{(\beta)}\right]^{\mathrm{D}}, \qquad K = 1, \ldots, 2d.$$

The matrix $U_K^{(\beta)}$ is anti-self-dual, i.e.:

$$U_K^{(\beta)} = -\big[U_K^{(\beta)}\big]^{\mathrm{D}}. \tag{9.26}$$

Moreover, they satisfy the following relations with P and R (we drop the superscript β from the matrix $Q^{(\beta)}$, $P^{(\beta)}$ and $R^{(\beta)}$ for simplicity) which are also known in the literature as a Lax-pair form:

$$\underbrace{\partial u_K P = \big[U_K^{(\beta)}, P\big], \qquad \partial u_K R = \big[U_K^{(\beta)}, R\big]}_{\text{Lax-pair form}}. \tag{9.27}$$

Explicitly, $U_K^{(\beta)}$ can be written as

$$U_K^{(\beta)} = -\frac{1}{K}\big((Q^K)_+ - (Q^K)_-^{\mathrm{D}}\big) - \frac{1}{2K}\big((Q^K)_0 - (Q^K)_0^{\mathrm{D}}\big), \tag{9.28}$$

where Q_+, Q_- denote the upper and lower triangular quaternion matrix while Q_0 is the diagonal quaternion.

PROOF. Differentiating (3.20) with respect to u_K and using (9.24) and (9.25), we get (9.26). Equation (9.27) follows by interchanging the operators $\partial/\partial x$ and $x\partial/\partial x$ with $\partial/\partial u_K$. Finally to prove (9.28), we start with the (2×1) normalized skew-orthogonal functions

$$\Phi_n^{(\beta)}(x) := \frac{1}{\sqrt{g_n^{(\beta)}}} \mathrm{e}^{-V(x)} \Pi_n(x), \qquad n \in \mathbb{N}, \tag{9.29}$$

where $V(x)$ is defined in (3.25) and

$$\Pi_n(x) = \begin{pmatrix} \pi_{2n}(x) \\ \pi_{2n+1}(x) \end{pmatrix} \tag{9.30}$$

is the monic skew-orthogonal polynomial. To save cluttering, we have suppressed the β dependence of $\pi_n(x)$. We have used $g_{2n}^{(\beta)} = g_{2n+1}^{(\beta)}$. Differentiating with respect to the deformation parameter, we get

$$\partial u_K \Phi_n^{(\beta)}(x) = -\frac{1}{2g_n^{(\beta)}} \partial u_K (g_n^{(\beta)}) \Phi_n^{(\beta)}(x) + \frac{\mathrm{e}^{-V(x)}}{\sqrt{g_n^{(\beta)}}} \partial u_K \Pi_n(x) - \frac{x^K}{K} \Phi_n^{(\beta)}(x). \tag{9.31}$$

Also, differentiating the skew-normalization condition, we get

$$\begin{aligned} -\frac{\partial}{\partial u_K} g_n^{(\beta)} &= -\int \big[\partial u_K (\Phi_n^{(\beta)}(x)) \widehat{\Psi}_n^{(\beta)}(x) + \Phi_n^{(\beta)}(x) \partial u_K \widehat{\Psi}_n^{(\beta)}(x)\big] \, \mathrm{d}x \\ &= \frac{1}{K}\big[Q^K + (Q^K)^{\mathrm{D}}\big]_{n,n} g_n^{(\beta)}. \end{aligned} \tag{9.32}$$

The diagonal elements have the form

$$\begin{aligned} \big(U_K^{(\beta)}\big)_{n,n} &= -\left(\frac{Q^K}{K}\right)_{n,n} + \frac{1}{2K}\big((Q^K)_{n,n} + (Q^K)_{n,n}^{\mathrm{D}}\big) \\ &= -\frac{1}{2K}\big[Q_{n,n}^K - (Q^K)_{n,n}^{\mathrm{D}}\big]. \end{aligned} \tag{9.33}$$

Thus, we get

$$U_K^{(\beta)} = -\frac{1}{K}\big(Q_+^K - (Q^K)_-^{\mathrm{D}}\big) - \frac{1}{2K}\big((Q_0^K) - (Q_0^K)^{\mathrm{D}}\big). \tag{9.34}$$

□

Remark. Here the operation $(Q^K)^{\mathrm{D}}_{-}$ means that the dual is taken first and then the lower triangular part collected. Also, due to the arbitrariness in the definition of an anti-self-dual matrix, we may choose the lower off-diagonal element of the diagonal quaternions in $U_K^{(\beta)}$ to be zero. With this choice, we can remove the arbitrariness in the definition of the odd skew-orthonormal polynomials.

The difference equation. In this section, we introduce a sequence of companion-like matrices $A_n^{(\beta)}(x)$ and $B_n^{(\beta)}(x)$ of sizes $2d \times 2d$ and show that the finite sub-sequence of skew-orthogonal vectors satisfy a system of difference equations, i.e.:

$$\begin{array}{c|c} \Phi^{(4)}_{W+1}(x) = A^{(4)}_N(x)\Phi^{(4)}_W(x), & \Phi^{(4)}_{W-1}(x) = \left(A^{(4)}_{N-1}(x)\right)^{-1}\Phi^{(4)}_W(x), \\ \hline \Psi^{(1)}_{W+1}(x) = A^{(1)}_N(x)\Psi^{(1)}_W(x), & \Psi^{(1)}_{W-1}(x) = \left(A^{(1)}_{N-1}(x)\right)^{-1}\Psi^{(1)}_W(x), \end{array} \tag{9.35}$$

and

$$\begin{array}{c|c} \Psi^{(4)}_{W+1}(x) = B^{(4)}_N(x)\Psi^{(4)}_W(x), & \Psi^{(4)}_{W-1}(x) = \left(B^{(4)}_{N-1}(x)\right)^{-1}\Psi^{(4)}_W(x), \\ \hline \Phi^{(1)}_{W+1}(x) = B^{(1)}_N(x)\Phi^{(1)}_W(x), & \Phi^{(1)}_{W-1}(x) = \left(B^{(1)}_{N-1}(x)\right)^{-1}\Phi^{(1)}_W(x), \end{array} \tag{9.36}$$

Using the relations in (9.5) and (9.6), we can write

$$\sum_{k=-d}^{d} \alpha_k^{(4)}(n)\Phi^{(4)}_{n-k}(x) = 0 \qquad \sum_{k=-d}^{d} \alpha_k^{(1)}(n)\Psi^{(1)}_{n-k}(x) = 0 \qquad \forall k \in N, \tag{9.37}$$

and

$$\begin{aligned} &\sum_{k=-d}^{d} \left(\alpha_k^{(1)}(n) - \delta_{0,k}\right)\Phi^{(1)}_{n-k}(x) = 0 \\ &\sum_{k=-d}^{d} \left(\alpha_k^{(4)}(n) - \delta_{0,k}\right)\Psi^{(4)}_{n-k}(x) = 0 \qquad \forall k \in \mathbb{N}, \end{aligned} \tag{9.38}$$

A quick glance at the generalized Christoffel – Darboux matrix (3.57) reveals that the relevant vectors contributing to the correlation function are $\phi^{(\beta)}_{2N-2d}(x), \ldots, \phi^{(\beta)}_{2N+2d-1}(x)$. This prompts us to define a finite subsequence (or window: $W \Rightarrow \{N-d, \ldots, N+d-1\}$, $N \geq d$) of skew-orthogonal vectors of size $2d$ (in the quaternion sense), $\Phi^{(\beta)}_W(x)$ and $\Psi^{(\beta)}_W(x)$:

$$\Phi^{(\beta)}_W(x) := \begin{pmatrix} \phi^{(\beta)}_{2N-2d}(x) \\ \vdots \\ \phi^{(\beta)}_{2N+2d-1}(x) \end{pmatrix} \equiv \begin{pmatrix} \Phi^{(\beta)}_{N-d}(x) \\ \vdots \\ \Phi^{(\beta)}_{N+d-1}(x) \end{pmatrix}, \qquad N \geq d, \tag{9.39}$$

$$\Psi^{(\beta)}_W(x) := \begin{pmatrix} \psi^{(\beta)}_{2N-2d}(x) \\ \vdots \\ \psi^{(\beta)}_{2N+2d-1}(x) \end{pmatrix} \equiv \begin{pmatrix} \Psi^{(\beta)}_{N-d}(x) \\ \vdots \\ \Psi^{(\beta)}_{N+d-1}(x) \end{pmatrix}, \qquad N \geq d, \tag{9.40}$$

such that

$$\Phi^{(\beta)}_{W\pm j}(x) := \left(\Phi^{(\beta)\mathrm{t}}_{N-d\pm j}(x)\cdots\Phi^{(\beta)\mathrm{t}}_{N+d-1\pm j}(x)\right)^{\mathrm{t}}, \qquad \forall j\in\mathbb{N}, \tag{9.41}$$

$$\Psi^{(\beta)}_{W\pm j}(x) := \left(\Psi^{(\beta)\mathrm{t}}_{N-d\pm j}(x)\cdots\Psi^{(\beta)\mathrm{t}}_{N+d-1\pm j}(x)\right)^{\mathrm{t}}, \qquad \forall j\in\mathbb{N}. \tag{9.42}$$

The rank of the window is equal to the degree of the potential $V(x)$ in the quaternion space and twice that of the degree of $V(x)$ in the real space.

We get for $N\geq d$:

$$\Phi^{(4)}_{W+1}(x) := \begin{pmatrix}\Phi^{(4)}_{N-d+1}(x)\\ \vdots\\ \Phi^{(4)}_{N+d}(x)\end{pmatrix} = A^{(4)}_n(x)\begin{pmatrix}\Phi^{(4)}_{N-d}(x)\\ \vdots\\ \Phi^{(4)}_{N+d-1}(x)\end{pmatrix} = A^{(4)}_n(x)\Phi^{(4)}_W(x), \tag{9.43}$$

and

$$\Psi^{(1)}_{W+1}(x) := \begin{pmatrix}\Psi^{(1)}_{N-d+1}(x)\\ \vdots\\ \Psi^{(1)}_{N+d}(x)\end{pmatrix} = A^{(1)}_n(x)\begin{pmatrix}\Psi^{(1)}_{N-d}(x)\\ \vdots\\ \Psi^{(1)}_{N+d-1}(x)\end{pmatrix} = A^{(1)}_n(x)\Psi^{(1)}_W(x), \tag{9.44}$$

where (suppressing the β dependence of $\alpha_j(n)$)

$$A^{(\beta)}_n(x) = \begin{pmatrix} 0 & \mathbf{1} & 0 & \cdots & 0\\ 0 & 0 & \mathbf{1} & \ddots & 0\\ \vdots & \vdots & \vdots & \vdots & \mathbf{1}\\ -(\alpha_{-d}(N))^{-1}\alpha_d(N) \ldots & -(\alpha_{-d}(N))^{-1}(\alpha_0(N)) \ldots & & & -(\alpha_{-d}(N))^{-1}\alpha_{-d+1}(N)\end{pmatrix}. \tag{9.45}$$

Remark. For potential $V(x)$ with even degree and $\forall n$, $\alpha_{\pm d}(n) = R^{(\beta)}_{n,n\pm d} - xP^{(\beta)}_{n,n\pm d}$ and its inverse have the quaternion structure

$$\alpha_{\pm d}(n) := \begin{pmatrix} a & 0\\ b & c\end{pmatrix}; \quad \alpha_{\pm d}(n)^{-1} = \begin{pmatrix} 1/a & 0\\ -b/ac & 1/c\end{pmatrix}, \qquad a,c\neq 0. \tag{9.46}$$

Hence they are invertible. This is the criterion for the existence of the skew-orthogonal polynomials corresponding to even potential.

Exercise. Show that for $V(x)$ to be a polynomial of order $2d+1$, $\alpha_{\pm d}(n)$ is noninvertible.

Using this, one can easily see that

$$\begin{aligned}\Phi^{(4)}_{W-1}(x) := \begin{pmatrix}\Phi^{(4)}_{N-d-1}(x)\\ \vdots\\ \Phi^{(4)}_{N+d-2}(x)\end{pmatrix} &= \left(A^{(4)}_{N-1}(x)\right)^{-1}\begin{pmatrix}\Phi^{(4)}_{N-d}(x)\\ \vdots\\ \Phi^{(4)}_{N+d-1}(x)\end{pmatrix}\\ &= \left(A^{(4)}_{N-1}(x)\right)^{-1}\Phi^{(4)}_W(x),\end{aligned} \tag{9.47}$$

and

$$\Psi^{(1)}_{W-1}(x) := \begin{pmatrix} \Psi^{(1)}_{N-d-1}(x) \\ \vdots \\ \Psi^{(1)}_{N+d-2}(x) \end{pmatrix} = \left(A^{(1)}_{N-1}(x)\right)^{-1} \begin{pmatrix} \Psi^{(1)}_{N-d}(x) \\ \vdots \\ \Psi^{(1)}_{N+d-1}(x) \end{pmatrix} = \left(A^{(1)}_{N-1}(x)\right)^{-1} \Psi^{(1)}_{W}(x), \tag{9.48}$$

where

$$\left(A^{(\beta)}_{N-1}(x)\right)^{-1} \tag{9.49}$$

$$= \begin{pmatrix} -\alpha_d^{-1}(N-1)\alpha_{d-1}(N-1) \ldots & -\alpha_d^{-1}(N-1)\alpha_0(N-1) \ldots & & -\alpha_d^{-1}(N-1)\alpha_{-d}(N-1) \\ \mathbf{1} & 0 & \cdots & \cdots & 0 \\ 0 & \ddots & \ddots & \ddots & 0 \\ \vdots & \vdots & \vdots & \mathbf{1} & 0 \end{pmatrix}.$$

Similarly, for $\Psi^{(4)}(x)$ and $\Phi^{(1)}(x)$, $A^{(\beta)}(x)$ is replaced by $B^{(\beta)}(x)$, where

$$B^{(\beta)}_n(x) \tag{9.50}$$

$$= \begin{pmatrix} 0 & \mathbf{1} & 0 & \cdots & 0 \\ 0 & 0 & \mathbf{1} & \ddots & 0 \\ \vdots & \vdots & \vdots & \vdots & \mathbf{1} \\ -(\alpha_{-d}(N))^{-1}\alpha_d(N) \ldots & -(\alpha_{-d}(N))^{-1}(\alpha_0(N)-\mathbf{1}) \ldots & & & -(\alpha_{-d}(N))^{-1}\alpha_{-d+1}(N) \end{pmatrix}.$$

More generally, this ladder operator can be used successively to obtain

$$\Phi^{(\beta)}_{W-j}(x) = \left(A^{(\beta)}_{N-j}(x)\right)^{-1} \cdots \left(A^{(\beta)}_{N-1}(x)\right)^{-1} \Phi^{(\beta)}_{W}(x), \qquad \forall j \in \mathbb{N}, \tag{9.51}$$

and

$$\Phi^{(\beta)}_{W+j}(x) = A^{(\beta)}_{N+j-1}(x) \cdots A^{(\beta)}_{n}(x) \Phi^{(\beta)}_{W}(x), \qquad \forall j \in \mathbb{N}. \tag{9.52}$$

This is the underlying idea behind folding, which is discussed in detail in the next section.

Folding.

Remark. From this point on we focus on the skew-orthogonal vectors $\Phi^{(4)}_n(x)$ (and their duals $\Psi^{(1)}_n(x)$), but everything being said can be immediately extended to the vectors $\Phi^{(1)}_n(x)$, (and hence $\Psi^{(4)}_n(x)$) by interchanging the rôles of the matrices $R^{(\beta)}$ by $R^{(\beta)} - \mathbf{1}$.

The notion of "folding" is the following: we express any skew-orthogonal function $\Phi^{(\beta)}_n(x)$ (or $\Psi^{(\beta)}_n(x)$) as a linear combination of $2d$ *fixed* consecutive vectors $\Phi^{(\beta)}_W(x) \equiv \Phi^{(4)}_{N-j}(x)$, $j = d, \ldots, -d+1$ *with polynomial coefficients.* We now provide a way of computing the folding function on the same line as done in [9] for bi-orthogonal polynomials.

For a fixed subsequence (or window) W, such that $W \Rightarrow \{N-d, \ldots, N+d-1\}$, $N \geq d$, we seek to describe the folding of the infinite wave-vector $\Phi^{(\beta)}(x)$ onto the window W by means of a single quaternion matrix $F(x)$ of size $\infty \times (2d)$ with polynomial entries such that:

$$\forall n, \qquad \Phi_n^{(\beta)}(x) = \sum_{k=N-d}^{N+d-1} F_{n,k}(x)\Phi_k^{(\beta)}(x). \tag{9.53}$$

(Similarly for $\Psi_n^{(\beta)}(x)$). In fact it is more convenient to think of $F(x)$ as a $\infty \times \infty$ matrix with only a vertical band of width $2d$ of nonzero entries (with column index in the range $N-d, \ldots, N+d-1$).

Now, we find an explicit form of the "folding function" in terms of some upper and lower triangular matrices. We derive the folding function for $\Phi^{(4)}(x)$ and $\Psi^{(1)}(x)$. For the folding function for $\Psi^{(4)}(x)$ and $\Phi^{(1)}(x)$ we follow the same procedure and hence state the result without proof. Note that everything is written in terms of quaternion matrices.

We will start with a few definitions and identities. We introduce the shift matrix Λ comprised of 2×2 blocks of unit matrix

$$\Lambda = \begin{pmatrix} 0 & \mathbf{1} & 0 & \ldots \\ 0 & 0 & \mathbf{1} & 0 \\ 0 & 0 & 0 & \ddots \\ \vdots & \vdots & \vdots & \ddots \end{pmatrix}, \qquad \mathbf{1} = \begin{pmatrix} 1 & 0 \\ 0 & 1 \end{pmatrix}, \tag{9.54}$$

satisfying the relation

$$\Lambda^{\mathrm{t}} = \Lambda^{\mathrm{D}}, \tag{9.55}$$

i.e., in this case, the transpose is the same as dual. Hence for convenience, for the rest of the chapter, we will denote it as Λ^{t}.

Let us define the projection matrices:

$$\overset{N}{\Pi} := \mathbf{1} - \underset{N}{\Pi}, \qquad \overset{N}{\underset{M}{\Pi}} := \underset{N}{\Pi} - \underset{M}{\Pi}. \tag{9.56}$$

We will also use the following identities,

$$\begin{aligned} &\overset{N+d}{\Pi}(\Lambda^{\mathrm{t}})^d = (\Lambda^{\mathrm{t}})^d \overset{N}{\Pi}, \qquad && \underset{N+d}{\Pi}(\Lambda^{\mathrm{t}})^d = (\Lambda^{\mathrm{t}})^d \underset{N}{\Pi}, \\ &\overset{N+d}{\Pi}\ \underset{N+d}{\Pi} = 0, && \overset{N+d}{\Pi}\ \underset{N}{\Pi} = 0, \end{aligned} \tag{9.57}$$

$$\begin{aligned} &\overset{N}{\Pi}\ \underset{N+d}{\Pi} = \underset{N+d}{\Pi} - \underset{N}{\Pi} = \overset{N+d}{\underset{N}{\Pi}}, \\ &(\Lambda^{\mathrm{t}})^d\Lambda^d = 1 - \underset{d}{\Pi}, \qquad \Lambda^d \underset{d}{\Pi} = 0. \end{aligned} \tag{9.58}$$

To find an explicit formula for the matrix $F(x)$ we will use the diagonal band matrices α_j, $j = -d, \ldots, d$ to express $R^{(\beta)} - xP^{(\beta)}$:

$$\overline{R}^{(\beta)}(x) = R^{(\beta)} - xP^{(\beta)} = \alpha_{-d}\Lambda^d + \sum_{k=1}^{d-1}\left[\alpha_{-k}\Lambda^k\right] + \sum_{k=0}^{d}\left[\alpha_k(\Lambda^{\mathrm{t}})^k\right], \tag{9.59}$$

where Λ is the shift matrix defined in (9.54). (For simplicity, we have again suppressed the β dependence of α_j.)

Using the shift matrix on (9.59), we get

$$(9.60)\quad (\Lambda^{\mathrm{t}})^d\alpha_{-d}^{-1}[\overline{R}^{(\beta)}(x)] = (\Lambda^{\mathrm{t}})^d(\Lambda)^d + (\Lambda^{\mathrm{t}})^d\alpha_{-d}^{-1}\sum_{k=1}^{d-2}\left[\alpha_{-k}\Lambda^k\right] + (\Lambda^{\mathrm{t}})^d\alpha_{-d}^{-1}\sum_{k=0}^{d}\left[\alpha_k(\Lambda^{\mathrm{t}})^k\right].$$

Thus we can define a semi-infinite matrix $G^{(\beta)}$, which is strictly lower triangular:

$$(9.61)\qquad G^{(\beta)} := 1 - \underset{d}{\Pi} - (\Lambda^{\mathrm{t}})^d\alpha_{-d}^{-1}[\overline{R}^{(\beta)}(x)].$$

Similarly

$$(9.62)\qquad \Lambda^d\alpha_d^{-1}[\overline{R}^{(\beta)}(x)] = 1 + \Lambda^d\alpha_d^{-1}\left[\sum_{k=1}^{d}\alpha_{-k}\Lambda^k + \sum_{k=0}^{d-1}\alpha_k(\Lambda^{\mathrm{t}})^k\right],$$

allows us to define a strictly upper triangular matrix

$$(9.63)\qquad C^{(\beta)} := 1 - \Lambda^d\alpha_d^{-1}\overline{R}^{(\beta)}(x).$$

Multiplying (9.61) by $\alpha_{-d}(\Lambda)^d$ on the left and $(1 - G^{(\beta)})^{-1}(\Lambda^{\mathrm{t}})^d\alpha_{-d}^{-1}$ on the right, we get

$$(9.64)\qquad \mathbf{1} = \overline{R}^{(\beta)}(x)\left[(1 - G^{(\beta)})^{-1}(\Lambda^{\mathrm{t}})^d\alpha_{-d}^{-1}\right],$$

while for the upper triangular matrix, we have the relation

$$(9.65)\qquad \mathbf{1} = (1 - C^{(\beta)})^{-1}\Lambda^d\alpha_d^{-1}\overline{R}^{(\beta)}(x).$$

Thus the matrix $[R^{(\beta)} - xP^{(\beta)}]$ has a left and right inverse, but they are not the same. One is upper triangular while the other is lower triangular. However, they satisfy the following relation:

$$(9.66)\qquad [\overline{R}^{(\beta)}(x)]_{\mathrm{L}}^{-1} = -([\overline{R}^{(\beta)}(x)]_{\mathrm{R}}^{-1})^{\mathrm{D}}.$$

One observes similar features for the matrix $(Q - x)$ in the context of the two-matrix model studied in [9].

Following the above procedure, we can also get the upper and lower triangular matrices, needed to fold $\Psi^{(4)}(x)$ and $\Phi^{(1)}(x)$. They are

$$(9.67)\quad \widetilde{G}^{(\beta)} := 1 - \underset{d}{\Pi} - (\Lambda^{\mathrm{t}})^d\alpha_{-d}^{-1}[\overline{R}^{(\beta)}(x) - \mathbf{1}], \qquad \widetilde{C}^{(\beta)} := 1 - \Lambda^d\alpha_d^{-1}[\overline{R}^{(\beta)}(x) - \mathbf{1}],$$

such that they satisfy

$$(9.68)\qquad \begin{aligned} \mathbf{1} &= \left[[\overline{R}^{(\beta)}(x) - \mathbf{1}]\right]\left[(1 - \widetilde{G}^{(\beta)})^{-1}(\Lambda^{\mathrm{t}})^d\alpha_{-d}^{-1}\right], \\ \mathbf{1} &= (1 - \widetilde{C}^{(\beta)})^{-1}\Lambda^d\alpha_d^{-1}[\overline{R}^{(\beta)}(x) - \mathbf{1}]. \end{aligned}$$

From the definition $[R^{(4)} - xP^{(4)}]\Phi^{(4)}(x) = 0$, and (9.61) and (9.63), we get

$$(9.69)\qquad \Phi^{(4)}(x) = C^{(4)}\Phi^{(4)}(x), \qquad \Phi^{(4)}(x) = \left(G^{(4)} + \underset{d+1}{\Pi}\right)\Phi^{(4)}(x).$$

Similarly from the definition $[R^{(1)} - xP^{(1)}]\Psi^{(1)}(x) = 0$, and (9.61) and (9.63), we get

$$(9.70)\qquad \Psi^{(1)}(x) = C^{(1)}\Psi^{(1)}(x), \qquad \Psi^{(1)}(x) = \left(G^{(1)} + \underset{d+1}{\Pi}\right)\Psi^{(1)}(x).$$

Given that the quaternion matrix G is strictly lower triangular, we can use (9.69) – (9.70) in an iterative manner to get

$$\begin{aligned} \overset{N}{\Pi}\,\Phi^{(4)}(x) &= (1-G^{(4)})^{-1}\overset{N}{\Pi}\,G^{(4)}\underset{N}{\Pi}\,\Phi^{(4)}(x), \\ \overset{N}{\Pi}\,\Psi^{(1)}(x) &= (1-G^{(1)})^{-1}\overset{N}{\Pi}\,G^{(1)}\underset{N}{\Pi}\,\Psi^{(1)}(x). \end{aligned} \tag{9.71}$$

We will drop the parameter β for simplicity. Also since the matrix G has d bands below the principal diagonal, we have for the folding function

$$\overset{\overline{N}}{\Pi}\,F(x) = (1-G)^{-1}\overset{\overline{N}}{\Pi}\,G\underset{\overline{N}}{\Pi}, \tag{9.72}$$

where $\overline{N} = N + d - 1$. Simplifying further, we replace G from (9.61) to get

$$\overset{\overline{N}}{\Pi}\,F(x) = (1-G)^{-1}\overset{\overline{N}}{\Pi}\Big[1 - \underset{d}{\Pi} - \alpha_{-d}^{-1}(\Lambda^{\mathrm{t}})^{d}(R^{(\beta)} - xP^{(\beta)})\Big]\underset{\overline{N}}{\Pi}. \tag{9.73}$$

Then using the identities obtained before, one gets

$$\begin{aligned} \overset{\overline{N}}{\Pi}\,F(x) &= -(1-G)^{-1}\overset{\overline{N}}{\Pi}\,\alpha_{-d}^{-1}(\Lambda^{\mathrm{t}})^{d}[\overline{R}^{(\beta)}(x)]\underset{\overline{N}}{\Pi}, \\ &= -\Big[(1-G)^{-1}\alpha_{-d}^{-1}(\Lambda^{\mathrm{t}})^{d}\overset{N}{\Pi}\big(\overline{R}^{(\beta)}(x)\big)\underset{N}{\Pi}\Big], \\ &= -\underset{N}{\overset{\overline{N}}{\Pi}} - (1-G)^{-1}\alpha_{-d}^{-1}(\Lambda^{\mathrm{t}})^{d}\Big[\overline{R}^{(\beta)}(x), \underset{N}{\Pi}\Big]. \end{aligned}$$

Thus we have for the upper folding

$$\overset{\overline{N}}{\Pi}\,F(x) = -\underset{N}{\overset{\overline{N}}{\Pi}} - (1-G)^{-1}\alpha_{-d}^{-1}(\Lambda^{\mathrm{t}})^{d}\Big[\overline{R}^{(\beta)}(x), \underset{N}{\Pi}\Big]. \tag{9.74}$$

Now, we will study the lower folding function. We can write

$$\begin{aligned} \underset{M}{\Pi}\,\Phi^{(4)}(x) &= (1-C^{(4)})^{-1}\underset{M}{\Pi}\,C^{(4)}\overset{M}{\Pi}\,\Phi^{(4)}(x), \\ \underset{M}{\Pi}\,\Psi^{(1)}(x) &= (1-C^{(1)})^{-1}\underset{M}{\Pi}\,C^{(1)}\overset{M}{\Pi}\,\Psi^{(1)}(x). \end{aligned} \tag{9.75}$$

where $M = N - d$. This relation is valid since we are folding finite band matrices. Thus $F(x)$ can be written as

$$\underset{M}{\Pi}\,F(x) = (1-C)^{-1}\underset{M}{\Pi}\,C\overset{M}{\Pi}. \tag{9.76}$$

Replacing C from (9.63) and using the identities, we get

$$\begin{aligned} \underset{M}{\Pi}\,F(x) &= -(1-C)^{-1}\Lambda^{d}\alpha_{d}^{-1}\underset{N}{\Pi}\,\overline{R}^{(\beta)}(x)\overset{M}{\Pi} \\ &= -\underset{M}{\overset{N}{\Pi}} + (1-C)^{-1}\Lambda^{d}\alpha_{d}^{-1}\Big[\overline{R}^{(\beta)}(x), \underset{N}{\Pi}\Big] \\ &= -\underset{M}{\overset{N}{\Pi}} + (1-C)^{-1}\Lambda^{d}\alpha_{d}^{-1}\Big[\overline{R}^{(\beta)}(x), \underset{N}{\Pi}\Big]. \end{aligned} \tag{9.77}$$

The folding function is

$$F(x) = \underset{M}{\Pi} F(x) + \overset{\overline{N}}{\Pi} F(x) + \underset{M}{\overset{N}{\Pi}}, \tag{9.78}$$

such that we get

$$F(x) = \left((1-C)^{-1}\Lambda^d \alpha_d^{-1} - (1-G)^{-1}\alpha_{-d}^{-1}(\Lambda^{\mathrm{t}})^d\right)\left[\overline{R}^{(\beta)}(x), \underset{N}{\Pi}\right]. \tag{9.79}$$

One can easily repeat the above calculations to find the folding function for $\Psi^{(4)}(x)$ and $\Phi^{(1)}(x)$. It gives

$$\widetilde{F}(x) = \left((1-\widetilde{C})^{-1}\Lambda^d \alpha_d^{-1} - (1-\widetilde{G})^{-1}\alpha_{-d}^{-1}(\Lambda^{\mathrm{t}})^d\right)\left[\overline{R}^{(\beta)}(x), \underset{N}{\Pi}\right]. \tag{9.80}$$

Remark. While studying bi-orthogonal polynomials, Bergère and Eynard [5] found a very elegant expression for the folding function. It would be interesting to see the existence of such compact expression for skew-orthogonal vectors.

Folded deformation matrices. Now we will show that a finite subsequence or window of these skew-orthonormal vectors satisfy a system of PDEs under the infinitesimal change of the deformation parameter. We prove this for $\Phi^{(4)}(x)$ and $\Psi^{(1)}(x)$. It can be trivially extended to $\Psi^{(4)}(x)$ and $\Phi^{(1)}(x)$. We show that these skew-orthogonal vectors satisfy a system of PDEs with respect to the deformation parameter which is given by

$$\begin{array}{c|c} \dfrac{\partial}{\partial u_K}\Psi_W^{(4)}(x) = \underline{U}_K^{N(4)}(x)\Psi_W^{(4)}(x), & \dfrac{\partial}{\partial u_K}\Phi_W^{(1)}(x) = \underline{U}_K^{N(1)}(x)\Phi_W^{(1)}(x), \\ \hline \dfrac{\partial}{\partial u_K}\Phi_W^{(4)}(x) = U_K^{N(4)}(x)\Phi_W^{(4)}(x), & \dfrac{\partial}{\partial u_K}\Psi_W^{(1)}(x) = U_K^{N(1)}(x)\Psi_W^{(1)}(x). \end{array} \tag{9.81}$$

where $\Phi_W^{(\beta)}(x)$ and $\Psi_W^{(\beta)}(x)$is defined in (9.39), (9.40).

We define the folded deformation matrix as

$$\underset{M}{\overset{\overline{N}}{\Pi}} U_K^{(\beta)} F(x) := U_K^{N(\beta)}, \quad \underset{M}{\overset{\overline{N}}{\Pi}} U_K^{(\beta)} \widetilde{F}(x) := \underline{U}_K^{N(\beta)}, \qquad K = 1, \dots 2d, \tag{9.82}$$

where we project a $2d \times \infty$ quaternion matrix $\Pi_M^{\overline{N}} U_K^{(\beta)}$ onto a $\infty \times 2d$ quaternion matrix $F(x)$ and $\widetilde{F}(x)$. We have

$$
\begin{aligned}
(9.83)\quad & -KU_K^{N(\beta)} \\
&= \mathop{\Pi}\limits_{M}^{\overline{N}}\left[[Q_+^K - (Q^K)_-^{\mathrm{D}}] + \tfrac{1}{2}[Q_0^K - (Q_0^K)^{\mathrm{D}}]\right]F(x) \\
&= \mathop{\Pi}\limits_{M}^{\overline{N}}\left[[Q_+^K - (Q^K)_-^{\mathrm{D}}] + \tfrac{1}{2}[Q_0^K - (Q_0^K)^{\mathrm{D}}]\right]\left[\mathop{\Pi}\limits_{M} F(x) + \mathop{\Pi}\limits^{\overline{N}} F(x) + \mathop{\Pi}\limits_{M}^{\overline{N}}\right] \\
&= -\mathop{\Pi}\limits_{M}^{\overline{N}}[KU_K^{(\beta)}]\mathop{\Pi}\limits_{M}^{\overline{N}} + \mathop{\Pi}\limits_{M}^{\overline{N}} Q^K \mathop{\Pi}\limits^{\overline{N}} F(x) - \mathop{\Pi}\limits_{M}^{\overline{N}}(Q^K)^D \mathop{\Pi}\limits_{M} F(x) \\
&= -\mathop{\Pi}\limits_{M}^{\overline{N}}[KU_K^{(\beta)}]\mathop{\Pi}\limits_{M}^{\overline{N}} - \mathop{\Pi}\limits_{M}^{\overline{N}} Q^K \left[\mathop{\Pi}\limits_{N}^{\overline{N}} + (1-G)^{-1}(\Lambda^{\mathrm{t}}\alpha_{-d}^{-1})^d\left[\overline{R}^{(\beta)}(x), \mathop{\Pi}\limits_{N}\right]\right] \\
&\qquad - \mathop{\Pi}\limits_{M}^{\overline{N}}(Q^K)^D\left[-\mathop{\Pi}\limits_{N}^{M} + (1-C)^{-1}(\alpha_{-d}^{-1}\Lambda)^d\left[\overline{R}^{(\beta)}(x), \mathop{\Pi}\limits_{N}\right]\right] \\
&= -\mathop{\Pi}\limits_{M}^{\overline{N}}[KU_K^{(\beta)}]\mathop{\Pi}\limits_{M}^{\overline{N}} - \mathop{\Pi}\limits_{M}^{\overline{N}} Q^K \mathop{\Pi}\limits_{N}^{\overline{N}} + \mathop{\Pi}\limits_{M}^{\overline{N}}(Q^K)^{\mathrm{D}}\mathop{\Pi}\limits_{N}^{M} \\
&\qquad - \mathop{\Pi}\limits_{M}^{\overline{N}}\left[(Q^K)(1-G)^{-1}(\Lambda^{\mathrm{t}}\alpha_{-d}^{-1})^d\left[\overline{R}^{(\beta)}(x), \mathop{\Pi}\limits_{N}\right]\right] \\
&\qquad - \mathop{\Pi}\limits_{M}^{\overline{N}}\left[(Q^K)^{\mathrm{D}}(1-C)^{-1}(\alpha_{-d}^{-1}\Lambda)^d\left[P^{(\beta)}, \mathop{\Pi}\limits_{N}\right]\right] \\
&= \mathop{\Pi}\limits_{M}^{N}\frac{(Q_0^K + (Q_0^K)^{\mathrm{D}})}{2}\mathop{\Pi}\limits_{M}^{N} - \mathop{\Pi}\limits_{N+1}^{\overline{N}}\frac{(Q_0^K + (Q_0^K)^{\mathrm{D}})}{2}\mathop{\Pi}\limits_{N+1}^{\overline{N}} - 2x^K\left[\mathop{\Pi}\limits_{M}^{N} - \mathop{\Pi}\limits_{N}^{\overline{N}}\right] \\
&\qquad - \mathop{\Pi}\limits_{N+1}^{\overline{N}}[Q^K - x^K]\left(\overline{R}^{(\beta)}(x)\right)_{\mathrm{R}}^{-1}\left[\overline{R}^{(\beta)}(x), \mathop{\Pi}\limits_{N}\right] \\
&\qquad - \mathop{\Pi}\limits_{M}^{\overline{N}}(Q^K - x^K)^{\mathrm{D}}\left(\overline{R}^{(\beta)}(x)\right)_{\mathrm{L}}^{-1}\left[\overline{R}^{(\beta)}(x), \mathop{\Pi}\limits_{N}\right] \\
&= \mathop{\Pi}\limits_{M}^{\overline{N}}\frac{(Q_0^K + (Q_0^K)^{\mathrm{D}})}{2}\left[\mathop{\Pi}\limits_{M}^{N} - \mathop{\Pi}\limits_{N}^{\overline{N}}\right] - \mathop{\Pi}\limits_{M}^{\overline{N}}\mathcal{W}_K(x)\left[\overline{R}^{(\beta)}(x), \mathop{\Pi}\limits_{N}\right] \\
&\qquad - 2x^K\left[\mathop{\Pi}\limits_{M}^{N} - \mathop{\Pi}\limits_{N}^{\overline{N}}\right],
\end{aligned}
$$

where

$$
(9.84)\qquad \mathcal{W}_K(x) = (Q^K - x^K)\left(\overline{R}^{(\beta)}(x)\right)_{\mathrm{R}}^{-1} + (Q^K - x^K)^{\mathrm{D}}\left(\overline{R}^{(\beta)}(x)\right)_{\mathrm{L}}^{-1}.
$$

Note. For convenience, we have dropped the superscript β from $Q^{(\beta)}$ as it has no importance on the deformation system.

Now, if we define in analogy to the Pauli matrix σ_3,

$$
(9.85)\qquad \Sigma_3 := \left[\mathop{\Pi}\limits_{M}^{N} - \mathop{\Pi}\limits_{N}^{\overline{N}}\right] = \mathrm{diag}(\underbrace{\mathbf{1}\cdots\mathbf{1}}_{d}\underbrace{\mathbf{-1}\cdots\mathbf{-1}}_{d}),
$$

then we may write

$$
(9.86)\qquad \boxed{-KU_K^{N(\beta)} = \mathop{\Pi}\limits_{M}^{\overline{N}}\frac{(Q_0^K + (Q_0^K)^{\mathrm{D}})}{2}\Sigma_3 - \mathop{\Pi}\limits_{M}^{\overline{N}}\mathcal{W}_K(x)\left[\overline{R}^{(\beta)}(x), \mathop{\Pi}\limits_{N}\right] - 2x^K\Sigma_3,}
$$

For $\Psi^{(4)}(x)$ and $\Phi^{(1)}(x)$, we have

$$\boxed{-K\underline{U}_K^{N(\beta)} = \prod_M^{\overline{N}} \frac{(Q_0^K + (Q_0^K)^{\mathrm{D}})}{2}\Sigma_3 - \prod_M^{\overline{N}} \underline{\mathcal{W}}_K(x)\Big[\overline{R}^{(\beta)}(x), \prod_N\Big] - 2x^K\Sigma_3,} \tag{9.87}$$

where

$$\underline{\mathcal{W}}_K(x) = (Q^K - x^K)(\overline{R}^{(\beta)}(x) - \mathbf{1})_{\mathrm{R}}^{-1} + (Q^K - x^K)^{\mathrm{D}}(\overline{R}^{(\beta)}(x) - \mathbf{1})_{\mathrm{L}}^{-1}. \tag{9.88}$$

The differential equation. Finally, using the results of the deformation equation, we derive the ODEs satisfied by the finite subsequence of skew-orthogonal vectors. We show that the vectors $\Phi_W^{(\beta)}(x)$ and $\Psi_W^{(\beta)}(x)$ satisfy a system of ODEs:

$$\begin{array}{c|c} \frac{\mathrm{d}}{\mathrm{d}x}\Psi_W^{(4)}(x) = \underline{D}_N^{(4)}(x)\Psi_W^{(4)}(x), & \frac{\mathrm{d}}{\mathrm{d}x}\Phi_W^{(1)}(x) = \underline{D}_N^{(1)}(x)\Phi_W^{(1)}(x), \\ \hline \frac{\mathrm{d}}{\mathrm{d}x}\Phi_W^{(4)}(x) = D_N^{(4)}(x)\Phi_W^{(4)}(x), & \frac{\mathrm{d}}{\mathrm{d}x}\Psi_W^{(1)}(x) = D_N^{(1)}(x)\Psi_W^{(1)}(x). \end{array} \tag{9.89}$$

Remark. We will calculate explicitly the differential operator for $\Phi^{(4)}(x)$ and $\Psi^{(4)}(x)$. The corresponding dual vectors $\Psi^{(1)}(x)$ and $\Phi^{(1)}(x)$ respectively can be obtained by a simple transfer of $Q^{(4)} \mapsto Q^{(1)}$ and hence $P^{(4)} \mapsto P^{(1)}$ and $R^{(4)} \mapsto R^{(1)}$. Everything else remains the same.

We recall that the differential operator on any of these vectors

$$\begin{aligned} \frac{\mathrm{d}}{\mathrm{d}x}\Phi^{(\beta)}(x) &= P^{(\beta)}\Phi^{(\beta)}(x) \\ &= -\Big[\big(V'(Q)\big)_+ - \big(V'(Q)\big)_-^{\mathrm{D}} + \tfrac{1}{2}\big[\big(V'(Q)\big)_0 - \big(V'(Q)\big)_0^{\mathrm{D}}\big]\Big]\Phi^{(\beta)}(x) \\ &= -\sum_{K=0}^{2d-1} u_{K+1}\big[Q_+^K - (Q^K)_-^{\mathrm{D}} + \tfrac{1}{2}[Q_0^K - (Q^K)_0^{\mathrm{D}}]\big]\Phi^{(\beta)}(x), \end{aligned} \tag{9.90}$$

and similarly for $\Psi^{(\beta)}(x)$. Since $D_N^{(\beta)}$ and $\underline{D}_N^{(\beta)}$ is the folded version of $P^{(\beta)}$, we observe that the differential equation satisfied by the finite subsequence of $\Phi^{(4)}(x)$, namely $\Phi_W^{(4)}(x)$, is given by

$$\frac{\mathrm{d}}{\mathrm{d}x}\Phi_W^{(4)}(x) = \prod_M^{\overline{N}} P^{(4)}F(x)\Phi^{(4)}(x) = D_N^{(4)}\Phi_W^{(4)}(x), \tag{9.91}$$

such that with $Q := Q^{(4)}$, $P := P^{(4)}$ and $R := R^{(4)}$, we have

$$\begin{aligned} D_N^{(4)} &= \prod_M^{\overline{N}} PF(x) \\ &= -\prod_M^{\overline{N}}\Big[\big(V'(Q)\big)_+ - \big(V'(Q)\big)_-^{\mathrm{D}} + \frac{1}{2}\Big[\big(V'(Q)\big)_0 - \big(V'(Q)\big)_0^{\mathrm{D}}\Big]\Big]F(x) \\ &= -\sum_{K=0}^{2d-1} u_{K+1}\prod_M^{\overline{N}}\Big[Q_+^K - (Q^K)_-^{\mathrm{D}} + \frac{1}{2}\big[Q_0^K - (Q^K)_0^{\mathrm{D}}\big]\Big]F(x) \\ &= -\sum_{K=0}^{2d-1} u_{K+1}(-KU_K^{N(4)}), \\ &= 2V'(x)\Sigma_3 + \prod_M^{\overline{N}} \mathcal{W}_K^N(x)\Big[\overline{R}(x), \prod_N\Big]. \end{aligned} \tag{9.92}$$

In the last step, we have used the anti-self-dual property of P.

Thus we have

$$\boxed{\begin{aligned}\frac{\mathrm{d}}{\mathrm{d}x}\Phi_W^{(4)}(x) &= \prod_M^{\overline{N}} PF(x)\Phi^{(4)}(x)\\ &= \left[2V'(x)\Sigma_3 + \prod_M^{\overline{N}} \mathcal{W}_K^N(x)\left[\overline{R}(x), \prod_N\right]\right]\Phi_W^{(4)}(x).\end{aligned}} \tag{9.93}$$

where

$$\mathcal{W}_K^N(x) = [V'(Q) - V'(x)](xP-R)_{\mathrm{R}}^{-1} + \left[(V'(Q)) - V'(x)\right]^{\mathrm{D}}(xP-R)_{\mathrm{L}}^{-1}. \tag{9.94}$$

Replacing $\Phi^{(4)}(x) \mapsto \Psi^{(1)}(x)$ and $Q^{(4)} \mapsto Q^{(1)}$, (and hence $P^{(4)} \mapsto P^{(1)}$ and $R^{(4)} \mapsto R^{(1)}$) we can obtain the differential equation for the dual vector $\Psi^{(1)}(x)$.

Explicitly, we have

$$D_N^{(4)} = 2V'(x)\begin{pmatrix} \mathbf{1} & 0 & 0 & \dots & 0 & 0\\ 0 & \ddots & 0 & 0 & 0 & 0\\ 0 & 0 & \mathbf{1} & \dots & 0 & 0\\ 0 & 0 & 0 & -\mathbf{1} & 0 & 0\\ 0 & 0 & 0 & 0 & \ddots & 0\\ 0 & 0 & 0 & 0 & 0 & -\mathbf{1}\end{pmatrix}_{2d\times 2d} \tag{9.95}$$
$$+\mathcal{W}_K^N(x)\begin{pmatrix} 0 & 0 & 0 & \alpha_{-d}(M) & 0 & 0\\ 0 & 0 & 0 & \vdots & \ddots & 0\\ 0 & 0 & 0 & \alpha_{-1}(N) & \dots & \alpha_{-d}(N)\\ -\alpha_d(N+1) & \dots & -\alpha_1(N+1) & 0 & 0 & 0\\ 0 & \ddots & \vdots & 0 & 0 & 0\\ 0 & 0 & -\alpha_d(\overline{N}) & 0 & 0 & 0\end{pmatrix}.$$

For $\Psi_W^{(4)}(x)$ we have

$$\boxed{\begin{aligned}\frac{\mathrm{d}}{\mathrm{d}x}\Psi_W^{(4)}(x) &= \prod_M^{\overline{N}} P\widetilde{F}(x)\Psi^{(4)}(x)\\ &= \left[2V'(x)\Sigma_3 + \prod_M^{\overline{N}} \underline{\mathcal{W}}_K^N(x)\left[\overline{R}(x), \prod_N\right]\right]\Psi_W^{(4)}(x),\end{aligned}} \tag{9.96}$$

where $\underline{\mathcal{W}}_K^N(x)$ is given by

$$\underline{\mathcal{W}}_K^N(x) = [V'(Q) - V'(x)](\overline{R}(x)-1)_{\mathrm{R}}^{-1} + \left[(V'(Q)) - V'(x)\right]^{\mathrm{D}}(\overline{R}(x)-1)_{\mathrm{L}}^{-1}. \tag{9.97}$$

For the dual vector $\Phi_W^{(1)}(x)$, we follow the same procedure. Replacing $\Psi^{(4)}(x) \mapsto \Phi^{(1)}(x)$ and $Q^{(4)} \mapsto Q^{(1)}$, (and hence $P^{(4)} \mapsto P^{(1)}$ and $R^{(4)} \mapsto R^{(1)}$) we can obtain the ODEs for the dual vector $\Phi_W^{(1)}(x)$.

3. Fundamental solutions

We have seen that a finite subsequence of (2×1) skew-orthogonal vectors $\Phi_{N-k}^{(\beta)}(x)$ and $\Psi_{N-k}^{(\beta)}(x)$, $k = -d+1, \dots, d$, satisfies a system of differential-difference-deformation equations. To obtain the fundamental solutions, useful in the Riemann–Hilbert analysis, we need to look for $2d$ other solutions. In this section, we will show that the Cauchy-like transforms of these skew-orthogonal vectors indeed form such solutions.

We look for some integral representations of these skew-orthogonal vectors as the possible "other solutions." From the previous experience in orthogonal [10] and bi-orthogonal polynomials [8], we look for some Cauchy-like transforms $\tilde{\Psi}_n^{(\beta)}(x)$ and $\tilde{\Phi}_n^{(\beta)}(x)$, and hope that at least for sufficiently large n, these functions form the simultaneous solutions for the system of difference-differential-deformation equations. The remaining solutions can be obtained by taking the Cauchy-like transforms of certain moment functions $f_j(x)$ of order j, to be defined below. This trick had been used previously to define Riemann–Hilbert problems for bi-orthogonal [50] and skew-orthogonal polynomials [69] respectively.

Let us define $f_j(x)$[1]:

$$f_j(x) := \exp[V(x)] \int_{\mathbb{R}} \epsilon(x-y) y^j \exp[-V(y)]\, \mathrm{d}y, \qquad j = 0, \dots, 2d-2. \tag{9.98}$$

We will show that the functions

$$\Psi_n^{(j,\beta)}(x) := \int_{\mathbb{R}} \frac{f_j(x) - f_j(z)}{x-z} \Psi_n^{(\beta)}(z) \exp[-V(z)]\, \mathrm{d}z, \qquad j = 0, \dots, 2d-2, \tag{9.99}$$

$$\Phi_n^{(j,\beta)}(x) := \int_{\mathbb{R}} \frac{f_j(x) - f_j(z)}{x-z} \Phi_n^{(\beta)}(z) \exp[-V(z)]\, \mathrm{d}z, \qquad j = 0, \dots, 2d-2, \tag{9.100}$$

and the Cauchy-like transform of the skew-orthogonal vectors:

$$\begin{aligned} \tilde{\Psi}_n^{(\beta)}(x) &= \exp[V(x)] \int_{\mathbb{R}} \frac{\Psi_n^{(\beta)}(z)\exp[-V(z)]}{(x-z)}\, \mathrm{d}z, \\ \tilde{\Phi}_n^{(\beta)}(x) &= \exp[V(x)] \int_{\mathbb{R}} \frac{\Phi_n^{(\beta)}(z)\exp[-V(z)]}{(x-z)}\, \mathrm{d}z, \qquad n \in \mathbb{N}, \end{aligned} \tag{9.101}$$

are simultaneous solutions of the differential-difference-deformation equations.

Componentwise, using the recursion relations (9.2)–(9.4), we have:

$$\tilde{\psi}_n^{(4)}(x) = \sum P_{n,m}^{(4)} \tilde{\phi}_m^{(4)}(x), \qquad \tilde{\phi}_n^{(1)}(x) = \sum P_{n,m}^{(1)} \tilde{\psi}_m^{(1)}(x). \tag{9.102}$$

We also get

$$\begin{aligned} &\frac{\mathrm{d}}{\mathrm{d}x} \tilde{\psi}_n^{(\beta)}(x) \\ &= \exp[V(x)] \int_{\mathbb{R}} \frac{V'(x) - V'(z)}{x-z} \psi_n^{(\beta)}(z) \exp[-V(z)]\, \mathrm{d}z + \sum P_{n,m}^{(\beta)} \tilde{\psi}_m^{(\beta)}(x) \\ &= \exp[V(x)] \sum_{m=0}^{2d-2} C_m^{(\beta)}(x) Z_{m,n} + \sum P_{n,m}^{(\beta)} \tilde{\psi}_m^{(\beta)}(x), \end{aligned} \tag{9.103}$$

[1] Here, we note that if we wish to expand $\psi_j^{(1)}(x)$ in a polynomial basis, then for $j = 0, \dots, 2d-2$, they remain undefined (see duality section for a proof). To define these functions, we introduce the above variables.

where $V'(x) = (\mathrm{d}/\,\mathrm{d}x)V(x)$ and $V'(z) = (\mathrm{d}/\,\mathrm{d}z)V(z)$. Here, we have used

$$\exp[-V(z)]\left[\frac{\{V'(x)-V'(z)\}}{(x-z)}\right] = \sum_{m=0}^{2d-2} C_m^{(\beta)}(x)\phi_m^{(\beta)}(z). \tag{9.104}$$

Thus we have

$$\frac{\mathrm{d}}{\mathrm{d}x}\tilde{\psi}_n^{(\beta)}(x) = \sum P_{n,m}^{(\beta)}\tilde{\psi}_m^{(\beta)}(x), \qquad n \geq 2d. \tag{9.105}$$

Using (9.102) and (9.105), we get

$$\begin{gathered} \tilde{\phi}_n'^{(4)}(x) = \tilde{\psi}_n^{(4)}(x) = \sum P_{n,m}^{(4)}\tilde{\phi}_m^{(4)}(x), \\ \tilde{\psi}_n'^{(1)}(x) = \tilde{\phi}_n^{(1)}(x), \qquad \tilde{\phi}_n'^{(1)}(x) = \sum P_{n,m}^{(1)}\tilde{\phi}_m^{(1)}(x), \qquad n \geq 2d. \end{gathered} \tag{9.106}$$

Thus for $n \geq 2d-1$, our chosen functions satisfy the same system of differential equations as done by the skew-orthogonal vectors $\phi_n^{(\beta)}(x)$ and $\psi_n^{(\beta)}(x)$ (9.2).

Having obtained the relations between the $\tilde{\phi}^{(\beta)}(x)$ and $\tilde{\psi}^{(\beta)}(x)$, we will now show that these functions also satisfy the same recursion relations. We start with

$$\begin{aligned} x\frac{\mathrm{d}}{\mathrm{d}x}\tilde{\psi}_n^{(\beta)}(x) &= x\frac{\mathrm{d}}{\mathrm{d}x}\exp[V(x)]\int_{\mathbb{R}}\frac{\psi_n^{(\beta)}(z)\exp[-V(z)]}{(x-z)}\,\mathrm{d}z \\ &= xV'(x)\tilde{\psi}_n^{(\beta)}(x) - \exp[V(x)]\int_{\mathbb{R}}\frac{xV'(z)}{x-z}\psi_n^{(\beta)}(z)\exp[-V(z)]\,\mathrm{d}z \\ &\quad + \exp[V(x)]\int_{\mathbb{R}}\frac{x\psi_n^{(\beta)\prime}(z)}{x-z}\exp[-V(z)]\,\mathrm{d}z. \end{aligned}$$

Now, replacing $x \to x - z + z$ on the right-hand side, we get

$$\begin{aligned} x\frac{\mathrm{d}}{\mathrm{d}x}\tilde{\psi}_n^{(\beta)}(x) = \exp[V(x)]&\int_{\mathbb{R}}\frac{xV'(x)-zV'(z)}{x-z}\psi_n^{(\beta)}(z)\exp[-V(z)]\,\mathrm{d}z \\ &+ \exp[V(x)]\int_{\mathbb{R}}\frac{\mathrm{d}}{\mathrm{d}z}\left(\psi_n^{(\beta)}(z)\exp[-V(z)]\right)\mathrm{d}z \\ &+ \exp[V(x)]\int_{\mathbb{R}}\frac{\exp[-V(z)]}{x-z}\left[z\frac{\mathrm{d}}{\mathrm{d}z}\psi_n^{(\beta)}(z)\right]\mathrm{d}z. \end{aligned} \tag{9.107}$$

Thus we have

$$\begin{aligned} &x\frac{\mathrm{d}}{\mathrm{d}x}\tilde{\psi}_n^{(\beta)}(x) \\ &= \begin{cases} \exp[V(x)]\displaystyle\sum_{m=0}^{2d-1} d_m^{(\beta)}(x)Z_{m,n} + \sum R_{nm}^{(\beta)}\tilde{\psi}_n^{(\beta)}(x), & \beta = 1, \\ \exp[V(x)]\displaystyle\sum_{m=0}^{2d-1} d_m^{(\beta)}(x)Z_{m,n} + \sum (R_{nm}^{(\beta)} - \delta_{n,m})\tilde{\psi}_n^{(\beta)}(x), & \beta = 4. \end{cases} \end{aligned} \tag{9.108}$$

Here, we have used

$$\exp[-V(z)]\left[\frac{\{xV'(x)-zV'(z)\}}{(x-z)}\right] = \sum_{m=0}^{2d-1} d_m^{(\beta)}(x)\phi_m^{(\beta)}(z). \tag{9.109}$$

Thus for $n \geq 2d$, we can see that the first term drops out and hence $\tilde{\psi}_n^{(\beta)}(x)$ satisfy (8.91), (9.4) respectively.

Similarly, for $\beta = 4$, we get for $n \geq 2d$,

$$x\frac{\mathrm{d}}{\mathrm{d}x}\tilde{\phi}_n^{(4)}(x) = x\tilde{\psi}_n^{(4)}(x) \tag{9.110}$$
$$= \exp[V(x)]\int_{\mathbb{R}} \frac{(x - z + z)}{x - z}\psi_n^{(4)}(z)\exp[-V(z)]\,\mathrm{d}z,$$
$$= \sum R_{n,m}^{(4)}\tilde{\phi}_m^{(4)}(x),$$

where the first term drops out due to the condition $n \geq 2d$, $d \geq 1$. For $\beta = 1$, we differentiate (9.108) to get

$$x\frac{\mathrm{d}}{\mathrm{d}x}\tilde{\phi}_n^{(1)}(x) = \sum (R_{nm}^{(1)} - \delta_{n,m})\tilde{\phi}_n^{(1)}(x), \qquad n \geq 2d. \tag{9.111}$$

This in turn proves that for $n \geq 2d$, the Cauchy-like transforms satisfy the same set of recursion relations (9.37) and (9.38) as done by the skew-orthogonal vectors.

Thus we prove that our function is compatible under a system of difference-differential equations, i.e., the solution of the system of differential equations is also a simultaneous solution of the system of difference equations. We show that these solutions are also compatible with the system of deformation relations, i.e., they are also solutions of the system of PDEs.

We can see that

$$\frac{\partial}{\partial u_K}\tilde{\psi}_n^{(\beta)}(x) \tag{9.112}$$
$$= \frac{\exp[V(x)]}{K}\int_{\mathbb{R}} \frac{x^K - z^K}{x - z}\psi_n^{(\beta)}(z)\exp[-V(z)]\,\mathrm{d}z + \sum (U_K^\beta)_{nm}\tilde{\psi}_m^{(\beta)}(x)$$
$$= \exp[V(x)]\sum_{m=0}^{k-1} a_m^{(\beta)}(x)Z_{m,n} + \sum (U_K^\beta)_{n,m}\tilde{\psi}_m^{(\beta)}(x).$$

where

$$\exp[-V(z)]\left[\frac{\{x^K - z^K\}}{(x - z)}\right] = \sum_{m=0}^{K-1} a_m^{(\beta)}(x)\phi_m^{(\beta)}(z). \tag{9.113}$$

Thus for $n \geq 2d$, $\tilde{\psi}_n^{(\beta)}(x)$ satisfy (9.25). Also, combining (9.106), (9.112), we can see that

$$\tilde{\phi}_n^{(\beta)}(x) = \sum (U_K^\beta)_{n,m}\tilde{\phi}_m^{(\beta)}(x), \qquad n \geq 2d. \tag{9.114}$$

Thus for $n \geq 2d$, the new functions $\tilde{\phi}_n^{(\beta)}(x)$ and $\tilde{\psi}_n^{(\beta)}(x)$ satisfy the system of differential-difference-deformation equations and hence are mutually compatible.

In this section, we show that $\Psi_n^{(j,\beta)}(x)$ and $\Phi_n^{(j,\beta)}(x)$, $j = 0, \ldots, 2d-2$, are also solutions to the system of difference-deformation-differential equations. For this, we will characterize the skew-orthogonal functions $\phi_k^{(1)}(x)$ and $\psi_k^{(4)}(x)$ (which are polynomials of degree k and $k + 2d - 1$ respectively) through a set of skew-orthogonality relations with respect to the function $f_j(x)$. The expressions for these auxiliary solutions were suggested for $\beta = 1$ by the common referee of [37] and [69] using expressions contained therein, and later extended to $\beta = 4$.

We start with the identity for a polynomial $\Pi_j(x)$ of order j as

$$\Pi_j(x) = \exp[V(x)]\frac{\mathrm{d}}{\mathrm{d}x}(x^{j-2d+1}\exp[-V(x)]) \tag{9.115}$$

There are $2k+1$ orthogonality conditions for the skew-orthogonal polynomials of order $2k$ and $2k+1$. Starting with

$$(9.116)\quad Z_{j+2d-1,2k} = \int_{\mathbb{R}} \psi_{2k}^{(\beta)} \Pi_{j+2d-1} \exp[-V(x)]\, \mathrm{d}x = 0, \qquad \forall j = 0, \dots, 2k-2d+1,$$

and using (9.115), we get the $2k-2d+2$ conditions for

$$(9.117)\quad \beta = 1, \qquad Z_{2k,j+2d-1} = \int_{\mathbb{R}} \phi_{2k}^{(1)}(x) x^j \exp[-V(x)]\, \mathrm{d}x = 0, \qquad \forall j = 0, \dots, 2k-2d+1,$$

and

$$(9.118)\quad \beta = 4, \qquad Z_{2k,j+2d-1} = \int_{\mathbb{R}} \psi_{2k}^{\prime(4)}(x) x^j \exp[-V(x)]\, \mathrm{d}x = 0, \qquad \forall j = 0, \dots, 2k-2d+1,$$

while for the remaining $2d-1$ conditions, one can have

$$(9.119)\quad Z_{2k,j} = \int_{\mathbb{R}} \int_{\mathbb{R}} \phi_{2k}^{(1)}(x) y^j \epsilon(x-y) \exp[-V(y)]\, \mathrm{d}x\, \mathrm{d}y = 0, \qquad j = 0, \dots, 2d-2,$$

$$(9.120)\quad Z_{j,2k} = \int_{\mathbb{R}} \int_{\mathbb{R}} \psi_{2k}^{(4)}(x) y^{j+2d-1} \epsilon(x-y) \exp[-V(y)]\, \mathrm{d}x\, \mathrm{d}y = 0, \qquad j = 0, \dots, 2d-2.$$

Thus we define the function $f_j(x)$

$$(9.121)\quad f_j(x) := \exp[V(x)] \int_{\mathbb{R}} \epsilon(x-y) y^j \exp[-V(y)]\, \mathrm{d}y, \qquad j = 0, \dots, 2d-2.$$

Remark. Here, we note that for $\beta = 4$, without loss of generality, $f_j(x)$ can also be defined as

$$(9.122)\quad f_j(x) := \exp[V(x)] \int y^j \exp[-V(y)]\, \mathrm{d}y + c, \qquad j = 0, \dots, 2d-2,$$

where c is a constant that can be fixed by the skew-orthonormalization condition.

Having defined $f_j(x)$, we show that the functions

$$(9.123)\quad \begin{aligned} \Psi_n^{(j,\beta)}(x) &:= \int_{\mathbb{R}} \frac{f_j(x) - f_j(z)}{x-z} \Psi_n^{(\beta)}(z) \exp[-V(z)]\, \mathrm{d}z, \\ \Phi_n^{(j,\beta)}(x) &:= \int_{\mathbb{R}} \frac{f_j(x) - f_j(z)}{x-z} \Phi_n^{(\beta)}(z) \exp[-V(z)]\, \mathrm{d}z, \end{aligned}$$

satisfy the same system of difference-differential-deformation equations.

Componentwise, using the recursion relations (9.2)–(9.4), we get

$$(9.124)\quad \psi_n^{(j,4)}(x) = \sum P_{n,m}^{(4)} \phi_m^{(j,4)}(x), \qquad \phi_n^{(j,1)}(x) = \sum P_{n,m}^{(1)} \psi_m^{(j,1)}(x).$$

We also get

$$
\begin{aligned}
(9.125)\quad & \frac{\mathrm{d}}{\mathrm{d}x}\psi_n^{(j,\beta)}(x) \\
&= f_j(x)\int_{\mathbb{R}} \frac{V'(x)-V'(z)}{x-z}\psi_n^{(\beta)}(z)\exp[-V(z)]\,\mathrm{d}z \\
&\qquad + \int_{\mathbb{R}} \frac{x^j-z^j}{x-z}\psi_n^{(\beta)}(z)\exp[-V(z)]dz \\
&\qquad + \int_{\mathbb{R}} \frac{f_j(x)-f_j(z)}{x-z}\psi_n'^{(\beta)}(z)\exp[-V(z)]\,\mathrm{d}z \\
&= f_j(x)\sum_{m=0}^{2d-2} C_m^{(\beta)}(x)Z_{m,n} + \sum_{m=0}^{j-1} a_m^{(\beta)}(x)Z_{m,n} + \sum P_{n,m}^{(\beta)}\psi_m^{(j,\beta)}(x).
\end{aligned}
$$

Thus we have

$$
(9.126)\qquad \frac{\mathrm{d}}{\mathrm{d}x}\psi_n^{(j,\beta)}(x) = \sum P_{n,m}^{(\beta)}\psi_m^{(j,\beta)}(x), \qquad n \geq 2d.
$$

Using (9.124) and (9.126), we get

$$
(9.127)\qquad
\begin{aligned}
&\phi_n'^{(j,4)}(x) = \psi_n^{(j,4)}(x) = \sum P_{n,m}^{(4)}\phi_m^{(j,4)}(x), \\
&\psi_n'^{(j,1)}(x) = \phi_n^{(j,1)}(x), \quad \phi_n'^{(j,1)}(x) = \sum P_{n,m}^{(1)}\phi_m^{(j,1)}(x), \qquad n \geq 2d.
\end{aligned}
$$

Thus for $n \geq 2d$, our chosen functions satisfy the same system of differential equations as done by the skew-orthogonal vectors $\phi_n^{(\beta)}(x)$ and $\psi_n^{(\beta)}(x)$ (9.2).

Having obtained the relations between the $\phi^{(j,\beta)}(x)$ and $\psi^{(j,\beta)}(x)$, we show that these functions also satisfy the same recursion relations. We start with

$$
\begin{aligned}
(9.128)\quad x\frac{\mathrm{d}}{\mathrm{d}x}\psi_n^{(j,\beta)}(x) &= f_j(x)\int_{\mathbb{R}} \frac{(xV'(x)-zV'(z))}{(x-z)}\psi_n^{(\beta)}(z)\exp[-V(z)]\,\mathrm{d}z \\
&\quad + \int_{\mathbb{R}} \frac{(x^{j+1}-z^{j+1})}{(x-z)}\psi_n^{(\beta)}(z)\exp[-V(z)]\,\mathrm{d}z \\
&\quad + \int_{R} \frac{\mathrm{d}}{\mathrm{d}z}\big(\psi_n^{(\beta)}(z)(f_j(x)-f_j(z))\exp[-V(z)]\big)\,\mathrm{d}z \\
&\quad + \int_{\mathbb{R}} z\psi_n'^{(\beta)}(z)\frac{f_j(x)-f_j(z)}{x-z}\exp[-V(z)]\,\mathrm{d}z,
\end{aligned}
$$

which for $\beta = 1$ gives

$$
\begin{aligned}
(9.129)\quad x\frac{\mathrm{d}}{\mathrm{d}x}\psi_n^{(j,1)}(x) &= f_j(x)\sum_{m=0}^{2d-1} d_m^{(1)}(x)Z_{m,n} + \sum_{m=0}^{j} a_m^{(1)}(x)Z_{m,n} \\
&\qquad + \sum R_{nm}^{(1)}\psi_n^{(j,1)}(x), \qquad j = 0,\ldots,2d-2,
\end{aligned}
$$

and for $\beta = 4$ gives

$$
\begin{aligned}
(9.130)\quad x\frac{\mathrm{d}}{\mathrm{d}x}\psi_n^{(j,4)}(x) &= f_j(x)\sum_{m=0}^{2d-1} d_m^{(4)}(x)Z_{m,n} + 2\sum_{m=0}^{j} a_m^{(4)}(x)Z_{m,n} \\
&\qquad + \sum (R_{nm}^{(4)} - \delta_{n,m})\psi_n^{(j,4)}(x), \qquad j = 0,\ldots,2d-2.
\end{aligned}
$$

Thus for $n \geq 2d$, we can see that the first two terms drop out and hence $\psi_n^{(j,\beta)}(x)$, like $\psi_n^{(\beta)}(x)$ satisfy (8.91) and (9.4) respectively.

Similarly, for $\beta = 4$, we get for $n \geq 2d$,

$$x\frac{\mathrm{d}}{\mathrm{d}x}\phi_n^{(j,4)}(x) = x\psi_n^{(j,4)}(x) = \sum R_{n,m}^{(4)}\phi_m^{(j,4)}(x), \tag{9.131}$$

where the first term drops out due to the condition $n \geq 2d$, $d \geq 1$. For $\beta = 1$, we differentiate (9.108) to get

$$x\frac{\mathrm{d}}{\mathrm{d}x}\phi_n^{(j,1)}(x) = \sum(R_{nm}^{(1)} - \delta_{n,m})\phi_n^{(j,1)}(x), \qquad n \geq 2d. \tag{9.132}$$

This in turn proves that for $n \geq 2d$, these functions satisfy the same set of recursion relations (9.37) and (9.38) as done by the skew-orthogonal vectors.

Thus we prove that our function is compatible under a system of difference-differential equations, i.e., the solution of the system of differential equations is also a simultaneous solution of the system of difference equations. We show that these solutions are also compatible with the system of deformation relations, i.e., they are also solutions of the system of PDEs.

We can see that

$$\begin{aligned}\frac{\partial}{\partial u_K}\psi_n^{(j,\beta)}(x) &= \frac{f_j(x)}{K}\int_{\mathbb{R}}\frac{x^K - z^K}{x - z}\psi_n^{(\beta)}(z)\exp[-V(z)]\,\mathrm{d}z + \sum(U_K^\beta)_{nm}\psi_m^{(j,\beta)}(x)\\ &= f_j(x)\sum_{m=0}^{k-1}a_m^{(\beta)}(x)Z_{m,n} + \sum(U_K^\beta)_{n,m}\psi_m^{(j,\beta)}(x).\end{aligned} \tag{9.133}$$

Thus for $n \geq 2d$, $\psi_n^{(j,\beta)}(x)$ satisfy (9.25). Also, combining (9.127), (9.133), we can see that

$$\phi_n^{(j,\beta)}(x) = \sum(U_K^\beta)_{n,m}\phi_m^{(j,\beta)}(x), \qquad n \geq 2d. \tag{9.134}$$

Remark. Here we encounter terms of the form $f_{j+K}(x)$, $j = 0, \ldots, 2d-2$ and $K = 1, \ldots, 2d$. For $(j+K) > 2d-2$, this term does not contribute, by definition. Also if we consider only $K = 2d$, i.e., the leading term in $V(x)$, no such problem arise.

Thus for $n \geq 2d$, the new functions $\phi_n^{(j,\beta)}(x)$ and $\psi_n^{(j,\beta)}(x)$ satisfy the system of differential-difference-deformation equations and hence are mutually compatible.

Explicitly, we have

$$\begin{aligned}\Psi_{[W]}^{(\beta)} &= \begin{pmatrix}\Psi_{N-d}^{(\beta)}(x) & \tilde{\Psi}_{N-d}^{(\beta)}(x) & \Psi_{N-d}^{(0,\beta)}(x) & \ldots & \Psi_{N-d}^{(2d-2,\beta)}(x)\\ \vdots & \vdots & \vdots & \vdots & \vdots\\ \Psi_{N+d-1}^{(\beta)}(x) & \tilde{\Psi}_{N+d-1}^{(\beta)}(x) & \Psi_{N+d-1}^{(0,\beta)}(x) & \ldots & \Psi_{N+d-1}^{(2d-2,\beta)}(x)\end{pmatrix},\\ \Phi_{[W]}^{(\beta)} &= \begin{pmatrix}\Phi_{N-d}^{(\beta)}(x) & \tilde{\Phi}_{N-d}^{(\beta)}(x) & \Phi_{N-d}^{(0,\beta)}(x) & \ldots & \Phi_{N-d}^{(2d-2,\beta)}(x)\\ \vdots & \vdots & \vdots & \vdots & \vdots\\ \Phi_{N+d-1}^{(\beta)}(x) & \tilde{\Phi}_{N+d-1}^{(\beta)}(x) & \Phi_{N+d-1}^{(0,\beta)}(x) & \ldots & \Phi_{N+d-1}^{(2d-2,\beta)}(x)\end{pmatrix}.\end{aligned} \tag{9.135}$$

Thus we have obtained the fundamental solution to the system of differential-difference-deformation equations, satisfied by the skew-orthogonal vectors. From here, it should be possible to define the Riemann–Hilbert problem for skew-orthogonal vectors. It has been obtained in [69] for $\beta = 1$ and can be extended to $\beta = 4$.

4. Compatibility

In the previous section, we have obtained the fundamental system of solutions for the differential-difference-deformation equation.

In this section, we will provide an alternate proof to show that the recursion relations, the linear differential equations and the deformation equations are compatible in the sense that they admit a basis of simultaneous solutions, provided the vectors form a certain algebra defined by the commutation relations (9.8) and (9.27). In other words, here we will not use the explicit form of the fundamental solution to prove the compatibility.

This is the same as saying that the shifts $W \mapsto W+1$ in (9.43), (9.44) respectively, are compatible as vector differential-difference systems. This means that there exists a sequence of fundamental matrix solutions $\mathbf{\Phi}^{(\beta)}_{[W]}(x)$ and $\mathbf{\Psi}^{(\beta)}_{[W]}(x)$ which are simultaneous solutions of

$$\mathbf{\Phi}^{(4)}_{[W+1]}(x) = A^{(4)}_N(x)\mathbf{\Phi}^{(4)}_{[W]}(x), \tag{9.136}$$

$$\frac{\mathrm{d}}{\mathrm{d}x}\mathbf{\Phi}^{(4)}_{[W]}(x) = D^{(4)}_N(x)\mathbf{\Phi}^{(4)}_{[W]}(x), \tag{9.137}$$

and

$$\mathbf{\Psi}^{(4)}_{[W+1]}(x) = B^{(4)}_N(x)\mathbf{\Psi}^{(4)}_{[W]}(x), \tag{9.138}$$

$$\frac{\mathrm{d}}{\mathrm{d}x}\mathbf{\Psi}^{(4)}_{[W]}(x) = \underline{D}^{(4)}_N(x)\mathbf{\Psi}^{(4)}_{[W]}(x), \tag{9.139}$$

respectively. The same result holds for $\beta = 1$. This means that there exists a sequence of fundamental matrix solutions satisfying the recursion and differential equations

$$\mathbf{\Psi}^{(1)}_{[W+1]}(x) = A^{(1)}_N(x)\mathbf{\Psi}^{(1)}_{[W]}(x), \tag{9.140}$$

$$\frac{\mathrm{d}}{\mathrm{d}x}\mathbf{\Psi}^{(1)}_{[W]}(x) = D^{(1)}_N(x)\mathbf{\Psi}^{(1)}_{[W]}(x), \tag{9.141}$$

and

$$\mathbf{\Phi}^{(1)}_{[W+1]}(x) = B^{(1)}_n(x)\mathbf{\Phi}^{(1)}_{[W]}(x), \tag{9.142}$$

$$\frac{\mathrm{d}}{\mathrm{d}x}\mathbf{\Phi}^{(1)}_{[W]}(x) = \underline{D}^{(1)}_N(x)\mathbf{\Phi}^{(1)}_{[W]}(x), \tag{9.143}$$

respectively.

Exercise. Following similar arguments, show that the differential equations are also compatible with the shift $W \mapsto W-1$ ((9.47), (9.48)).

A similar procedure will be repeated to show the compatibility between the difference-deformation equations.

Difference-differential equations.

PROOF. Now we prove the compatibility of the differential-difference equations for $\Phi^{(4)}(x)$. The others are exactly similar and hence not repeated. The proof follows on similar line to the one outlined for bi-orthogonal polynomials [6].

Let us define a finite subsequence (or window) containing functions:

$$\bar{\Phi}_W^{(4)}(x) := \left[\bar{\Phi}_{N-d}^{(4)\mathrm{t}}(x), \ldots, \bar{\Phi}_{N+d-1}^{(4)\mathrm{t}}(x)\right]^{\mathrm{t}}, \qquad N \geq d, \tag{9.144}$$

which is a solution to the differential equation

$$\frac{\mathrm{d}}{\mathrm{d}x}\bar{\Phi}_W^{(4)}(x) = D_N^{(4)}(x)\bar{\Phi}_W^{(4)}(x). \tag{9.145}$$

A glance through the window will confirm that the shift $W \mapsto W+1$ will introduce one new function $\bar{\Phi}_{N+d}^{(4)}(x)$ inside the shifted window. Componentwise, let us suppose that they satisfy the recursion relation

$$\begin{aligned}[\eta_{-d}(m) - x\zeta_{-d}(m)]\bar{\Phi}_{m+d}^{(4)}(x)&\\ = x\sum_{l=-d+1}^{d} \zeta_l(m)\bar{\Phi}_{m-l}^{(4)}(x) - &\sum_{l=-d+1}^{d} \eta_l(m)\bar{\Phi}_{m-l}^{(4)}(x), \qquad m \geq d.\end{aligned} \tag{9.146}$$

It is easy to see that the differential equation componentwise reads

$$\frac{\mathrm{d}}{\mathrm{d}x}\bar{\Phi}_n^{(4)}(x) = \sum_{l=-d}^{d} \zeta_j(n)\bar{\Phi}_{n-l}^{(4)}(x), \qquad n = N-d, \ldots N+d-1, \tag{9.147}$$

where the $\bar{\Phi}_m^{(4)}(x)$ outside the window can be expressed in terms of that within the window by using (9.146) recursively. To show compatibility between the shift and differential operator, we need to show that the newly defined function, which is assumed to satisfy

$$\bar{\Phi}_{N+d}^{(4)}(x) = \big(\eta_{-d}(N)\big)^{-1}\left[x\sum_{l=-d}^{d} \zeta_l(N)\bar{\Phi}_{N-l}^{(4)}(x) - \sum_{l=-d+1}^{d} \eta_l(N)\bar{\Phi}_{N-l}^{(4)}(x)\right], \tag{9.148}$$

will also satisfy the same differential equation, i.e., we must show that the newly defined function $\bar{\Phi}_{N+d}^{(4)}(x)$ will also be a solution to the differential equation

$$\frac{\mathrm{d}}{\mathrm{d}x}\bar{\Phi}_{N+d}^{(4)}(x) = \sum_{l=-d}^{d} \zeta_j(N+d)\bar{\Phi}_{N-l+d}^{(4)}(x). \tag{9.149}$$

Remark. One must be careful about the position of the inverse of the quaternion, as in general, they do not commute.

Once this is proved, we can argue by induction that $\bar{\Phi}^{(4)}_{N+d+j}(x)$ satisfy the same sort of differential equation for any $j > 1$. To do this, we compute

$$\begin{aligned}(9.150)\quad \eta_{-d}(N)\left(\frac{\mathrm{d}}{\mathrm{d}x}\right)\bar{\Phi}^{(4)}_{N+d}(x) &= \frac{\mathrm{d}}{\mathrm{d}x}\left(x\sum_{l=-d}^{d}\zeta_l(N)\bar{\Phi}^{(4)}_{N-l}(x) - \sum_{l=-d+1}^{d}\eta_l(N)\bar{\Phi}^{(4)}_{N-l}(x)\right)\\ &= \sum_{l=-d}^{d}\zeta_l(N)\bar{\Phi}^{(4)}_{N-l}(x) + \sum_{l=-d}^{d}\sum_{j=-d}^{d}\zeta_l(N)\eta_j(N-l)\bar{\Phi}^{(4)}_{N-l-j}(x)\\ &\qquad - \sum_{l=-d}^{d}\sum_{j=-d}^{d}\eta_l(N)\zeta_j(N-l)\bar{\Phi}^{(1)}_{N-l-j}(x)\\ &\qquad + \eta_{-d}(N)\sum_{l=-d}^{d}\zeta_l(N+d)\bar{\Phi}^{(4)}_{N-l+d}(x).\end{aligned}$$

From (9.150), we can see that in order to prove (9.149), we will have to show that

$$\begin{aligned}(9.151)\quad \sum_{l=-d}^{d}\zeta_l\bar{\Phi}^{(4)}_{N-l}(x) &+ \sum_{l=-d}^{d}\sum_{j=-d}^{d}\zeta_l(N)\eta_j(N-l)\bar{\Phi}^{(4)}_{N-l-j}(x)\\ &- \sum_{l=-d}^{d}\sum_{j=-d}^{d}\eta_l(N)\zeta_j(N-l)\bar{\Phi}^{(4)}_{N-l-j}(x) = 0.\end{aligned}$$

But this is nothing but a direct consequence of (9.23), in terms of its components. We can repeat the same procedure for the shift $W \mapsto W-1$. Finally, one can extend this argument by induction to $\bar{\Phi}^{(4)}_{N+r}(x)$, thereby completing the proof. □

Exercise. Repeat the same procedure to the sequences

$$(9.152)\qquad \bar{\Psi}^{(1)}_W(x) := [\bar{\Psi}^{(1)\mathrm{t}}_{N-d}(x),\dots,\bar{\Psi}^{(1)\mathrm{t}}_{N+d-1}(x)]^{\mathrm{t}},\qquad N\geq d,$$

and with some minor modifications to

$$(9.153)\qquad \bar{\Phi}^{(1)}_W(x) := [\bar{\Phi}^{(1)\mathrm{t}}_{N-d}(x),\dots,\bar{\Phi}^{(1)\mathrm{t}}_{N+d-1}(x)]^{\mathrm{t}},\qquad N\geq d,$$

and

$$(9.154)\qquad \bar{\Psi}^{(4)}_W(x) := [\bar{\Psi}^{(4)\mathrm{t}}_{N-d}(x),\dots,\bar{\Psi}^{(4)\mathrm{t}}_{N+d-1}(x)]^{\mathrm{t}},\qquad N\geq d,$$

to prove compatibility conditions between the difference-differential equations satisfied by the skew-orthonormal vectors.

Difference-deformation equations.

PROOF. We show that these functions are also compatible with the deformation equations. We briefly outline the proof. We start with the same finite subsequence (or window) containing functions

$$(9.155)\qquad \bar{\Phi}^{(4)}_W(x) := [\bar{\Phi}^{(4)}_{N-d}(x),\dots,\bar{\Phi}^{(4)}_{N+d-1}(x)]^{\mathrm{t}},\qquad N\geq d,$$

which componentwise, will satisfy the PDE:

$$(9.156)\qquad \frac{\partial}{\partial u_K}\bar{\Phi}^{(4)}_n(x) = \sum_{j=0}^{K}U^K_j(n)\bar{\Phi}^{(4)}_{n-j}(x)\qquad n = N-d,\dots,N+d-1.$$

Then under a shift $W \mapsto W+1$, we get the new function $\bar{\Phi}^{(4)}_{N+d}(x)$ defined by

$$(\eta_{-d}(N))\bar{\Phi}^{(4)}_{N+d}(x) = x\left[\sum_{l=-d}^{d} \zeta_l(N)\bar{\Phi}_{N-d+1}(x) - \sum_{l=-d+1}^{d} \eta_l(N)\bar{\Phi}^{(4)}_{N-l}(x)\right], \tag{9.157}$$

which should satisfy

$$\frac{\partial}{\partial u_K}\bar{\Phi}^{(4)}_{N+d}(x) = \sum_{j=0}^{K} U_j^K(N+d)\bar{\Phi}^{(4)}_{N+d-j}(x). \tag{9.158}$$

To do this, we compute

$$\begin{aligned}
&\left(\frac{\partial}{\partial u_K}\right)(\eta_{-d}(N))\bar{\Phi}^{(4)}_{N+d}(x) \\
&= \frac{\partial}{\partial u_K}\left(x\left[\sum_{l=-d}^{d} \zeta_l(N)\bar{\Phi}^{(4)}_{N-l}(x) - \sum_{l=-d+1}^{d} \eta_l(N)\bar{\Phi}^{(4)}_{N-l}(x)\right]\right) \\
&= x\sum_{l=-d}^{d} \zeta_l'(N)\bar{\Phi}^{(4)}_{N-l}(x) + x\sum_{l=-d}^{d}\sum_{j=0}^{K} \zeta_l(N)U_j^K(N-l)\bar{\Phi}^{(4)}_{N-j-l}(x) \\
&\quad - \sum_{j=-d}^{d} \eta_l'(N)\bar{\Phi}^{(4)}_{N-l}(x) - \sum_{l=-d}^{d}\sum_{j=0}^{K} \eta_l(N)U_j^K(N-l)\bar{\Phi}^{(4)}_{N-j-l}(x) \\
&\quad + \eta_{-d}(N)\sum_{j=0}^{K} U_j^K(N+d)\bar{\Phi}^{(4)}_{N+d-j}(x) + \eta_{-d}'(N)\bar{\Phi}^{(4)}_{N+d}(x),
\end{aligned} \tag{9.159}$$

where $'$ denotes $\partial/\partial u_K$. Rearranging the coefficients in (9.149), we can see that basically we show that

$$\begin{aligned}
0 &= x\sum_{l=-d}^{d} \zeta_l'(N)\bar{\Phi}^{(4)}_{N-l}(x) + x\sum_{l=-d}^{d}\sum_{j=0}^{K} \zeta_l(N)U_j^K(N-l)\bar{\Phi}^{(4)}_{N-j-l}(x) \\
&\quad - \sum_{j=-d}^{d} \eta_l'(N)\bar{\Phi}^{(4)}_{N-l}(x) - \sum_{l=-d}^{d}\sum_{j=0}^{K} \eta_l(N)U_j^K(N-l)\bar{\Phi}^{(4)}_{N-j-l}(x).
\end{aligned} \tag{9.160}$$

But this is nothing but a direct consequence of the string (9.27), in terms of its components. Hence we get (9.158).

We can repeat the same procedure for the shift $W \mapsto W-1$. Finally, one can extend this argument by induction to $\bar{\Phi}^{(4)}_{N+r}(x)$, thereby completing the proof. □

Exercise. Repeat the same procedure to the sequences $\bar{\Psi}^{(1)}_W(x)$ and with some minor modifications to $\bar{\Phi}^{(1)}_W(x)$ and $\bar{\Psi}^{(4)}_W(x)$ to prove compatibility conditions between the difference-deformation equations satisfied by the skew-orthonormal vectors.

CHAPTER 10

Matrix Integral Representations and Zeros of Polynomials

We now turn to the integral representations for the three types of polynomials, namely the orthogonal and the two families of skew-orthogonal polynomials. These are given as averages over matrix ensembles. Results for orthogonal polynomials have been known for years, while those for skew-orthogonal polynomials were discovered independently by Eynard [30] and Ghosh [35]. We define the average of F with respect to the eigenvalue distribution (1.12) for given β, N as

$$\begin{aligned}(10.1)\quad &\langle F(x_1,\ldots,x_N)\rangle_{\beta,N}\\ &\qquad = \int\cdots\int F(x_1,\ldots,x_N)P_{\beta,N}(x_1,\ldots,x_N)\,\mathrm{d}x_1\cdots\,\mathrm{d}x_N.\end{aligned}$$

We consider monic polynomials, i.e., polynomials with highest coefficient unity. Then the orthogonal polynomials, appropriate for $\beta = 2$, are known [77] and are given by

$$P_j(x) = \left\langle \prod_{k=1}^{j}(x-x_k)\right\rangle_{2,j}. \tag{10.2}$$

The skew-orthogonal polynomials, appropriate for $\beta = 1$ (even-dimensional case), are given by

$$\pi^{(1)}_{2m}(x) = \left\langle \prod_{k=1}^{2m}(x-x_k)\right\rangle_{1,2m}, \tag{10.3}$$

$$\pi^{(1)}_{2m+1}(x) = \left\langle \left(x+\sum_{k=1}^{2m}x_k\right)\prod_{k=1}^{2m}(x-x_k)\right\rangle_{1,2m}. \tag{10.4}$$

For $\beta = 1$ (odd-dimensional case), we have

$$\pi^{(1)}_{2m}(x) = \left\langle \prod_{k=1}^{2m+1}(x-x_k)\right\rangle_{1,2m+1} - C_m, \tag{10.5}$$

$$\pi^{(1)}_{2m+1}(x) = \left\langle (x+\sum_{k=1}^{2m+1}x_k)\prod_{k=1}^{2m+1}(x-x_k)\right\rangle_{1,2m+1}, \tag{10.6}$$

valid for $m = 0, 1, \ldots, (N-3)/2$. Here C_m is determined from the condition (7.1). $\pi^{(1)}_{N-1}$ is given (in the nonmonic form) by

$$\pi^{(1)}_{N-1}(x) = \frac{\langle\prod_{k=1}^{N-1}(x-x_k)\rangle_{1,N-1}}{\int \mathrm{d}x\; w(x)\langle\prod_{k=1}^{N-1}(x-x_k)\rangle_{1,N-1}}. \tag{10.7}$$

For $\beta = 4$, the skew-orthogonal polynomials are

$$\pi_{2m}^{(4)}(x) = \left\langle \prod_{k=1}^{m} (x - x_k)^2 \right\rangle_{4,2m}, \tag{10.8}$$

$$\pi_{2m+1}^{(4)}(x) = \left\langle \left(x + 2\sum_{k=1}^{m} x_k \right) \prod_{k=1}^{m} (x - x_k)^2 \right\rangle_{4,2m}. \tag{10.9}$$

A proof of these integral representations is outlined in Appendix C (also see [35, 40]). Note that the above averages for all three β can be written in terms of $\det(x - H)$ and $(\operatorname{tr} H)\det(x - H)$ averaged over the matrix ensemble of H. We mention moreover that, Eynard has also obtained independently (10.3), (10.4) and (10.8), (10.9) and used the method of resolvents to calculate the asymptotics of skew-orthogonal polynomials with single support [30].

We finally talk briefly (and also nonrigorously) about the density of zeros of skew-orthogonal polynomials and the level densities of the corresponding ensembles. It is known that the orthogonal polynomials $P_j(x)$ have

(1) zeros which are real, simple and lie in the support of the weight functions.
(2) The zeros interlace for successive polynomials.

Numerical studies indicate that the zeros of the skew-orthogonal polynomials $\pi_{2m}^{(1)}(x), \pi_{2m+1}^{(1)}(x)$ and $\pi_{2m+1}^{(4)}(x)$ are also real, while those of the $\pi_{2m}^{(4)}(x)$ are complex, having small imaginary parts for large m and hence doubly degenerate real parts. To find the density of zeros, we take the eigenvalue spectrum $\{x_k\}$ to be ordered (i.e., $x_1 \leq x_2 \leq x_3 \leq \cdots$) and write $x_k = \langle x_k \rangle + \delta x_k$. The spectrum of the average eigenvalues $\{\langle x_k \rangle\}$ has asymptotically the density $R_1(x)$. The fluctuations δx_k are small [16], spanning over a few spacings (i.e., $\langle \delta x_k \delta x_l \rangle \approx (R_1(x))^{-2}$ for k, l not too far apart). Thus, for $\beta = 2$, $P_j(x) \simeq \Pi(x - \langle x_j \rangle)$, giving thereby the zeros as $\langle x_j \rangle$ with the density $R_1(x, j)$. For $\beta = 1$, we have similarly the density $R_1(x, 2m)$ for the zeros of $\pi_{2m}^{(1)}(x)$ with an additional term for the zero at $\sum \langle x_k \rangle$ for $\pi_{2m+1}^{(1)}(x)$. For $\beta = 4$, the leading approximations for $\pi_{2m}^{(4)}(x)$ in (10.8) gives doubly-degenerate zeros at $\langle x_k \rangle$ with density $2R_1(x, m)$, while $\pi_{2m+1}^{(4)}(x)$ has an additional term corresponding to the zero at $-2\sum \langle x_k \rangle$. Since $\pi_{2m}^{(4)}(x)$ is positive definite, the zeros of $\pi_{2m}^{(4)}(x)$ must be complex with small imaginary parts for large m.

CHAPTER 11

Duality

The duality between β and $4/\beta$ ensembles of random matrices have been of interest in recent times for all $\beta > 0$ after it was first studied by Mehta, Normand, Brézin and Hikami (see [14, 15, 55] and references therein), and later extended by [34] and [22]. The investigations were primarily based on finding the existence of duality for averages of products/ratios of characteristic polynomials.

More specifically, Mulase and Waldron [60], via an asymptotic expansion for quaternionic self-adjoint matrix integrals, showed that the $2N \times 2N$ Gaussian Orthogonal Ensemble (GOE) and $N \times N$ Gaussian Symplectic Ensemble (GSE) have exactly the same expansion term by term, except for a sign which appears from the contributions from graphs on a non-orientable surface with odd Euler characteristic. Later Bryc and Pierce [17] showed using combinatorics, the existence of duality between moments of the Gaussian Symplectic Ensemble and the Gaussian Orthogonal Ensemble, and between real Wishart and quaternionic Wishart ensembles.

In this chapter, we show the existence of duality between the two families of skew-orthogonal functions for a wide class of the orthogonal and symplectic ensembles of random matrices. This leads to the existence of duality relations between the generalized Christoffel – Darboux kernels and the systems of differential equations.

We select ensembles with classical and quartic double well potential and show that there indeed exists a duality between the two families of skew-orthogonal functions and consequently, their normalization constants.

The generalized Christoffel – Darboux formulas, arising in the study of ensembles of $N \times N$ self-dual quaternionic matrices, invariant under symplectic transformations, and $2N \times 2N$ real symmetric matrices, invariant under orthogonal transformations, are given by

$$
\begin{aligned}
S_{2N}^{(4)}(x,y) &= \frac{\widehat{\Phi}^{(4)}(x)[\overline{R}^{(4)}(x), \Pi_{2N}]\Phi^{(4)}(y)}{f(y)(y-x)}, \\
S_{2N}^{(1)}(x,y) &= \frac{\widehat{\Psi}^{(1)}(x)[\overline{R}^{(1)}(y), \Pi_{2N}]\Psi^{(1)}(y)}{f(x)(x-y)}, \qquad N \geq d,
\end{aligned}
\tag{11.1}
$$

where $\Phi^{(\beta)}(x)$ and $\Psi^{(\beta)}(x)$ are defined in (3.16) – (3.18) respectively. A look at (11.1) shows that duality can be observed between the statistical properties of these ensembles if we can show that there is a mapping between $\Phi^{(4)}(x) \mapsto \Psi^{(1)}(x)$, their derivatives, and $\overline{R}^{(4)}(x)/f(y) \mapsto -\overline{R}^{(1)}(y)/f(x)$.

Duality in differential system. In Chapter 9, we showed that a finite subsequence of the two families of skew-orthogonal functions with polynomial weight satisfy systems of differential-difference-deformation equations. We discuss in brief that duality relations also exist among the differential systems for skew-orthogonal functions corresponding to $\beta = 1$ and 4.

The recursion relations connecting the finite subsequence (or window) with the upper or lower one is through the ladder operator

$$\Phi^{(4)}_{W+1}(x) = A^{(4)}_N(x)\Phi^{(4)}_W(x), \qquad \Phi^{(4)}_{W-1}(x) = \left(A^{(4)}_{N-1}(x)\right)^{-1}\Phi^{(4)}_W(x), \tag{11.2}$$

$$\Psi^{(1)}_{W+1}(x) = A^{(1)}_N(x)\Psi^{(1)}_W(x), \qquad \Psi^{(1)}_{W-1}(x) = \left(A^{(1)}_{N-1}(x)\right)^{-1}\Psi^{(1)}_W(x), \tag{11.3}$$

and

$$\Psi^{(4)}_{W+1}(x) = B^{(4)}_N(x)\Psi^{(4)}_W(x), \qquad \Psi^{(4)}_{W-1}(x) = \left(B^{(4)}_{N-1}(x)\right)^{-1}\Psi^{(4)}_W(x), \tag{11.4}$$

$$\Phi^{(1)}_{W+1}(x) = B^{(1)}_N(x)\Phi^{(1)}_W(x), \qquad \Phi^{(1)}_{W-1}(x) = \left(B^{(1)}_{N-1}(x)\right)^{-1}\Phi^{(1)}_W(x), \tag{11.5}$$

The vectors $\Phi^{(\beta)}_W(x)$ and $\Psi^{(\beta)}_W(x)$ also satisfy a system of ODEs:

$$\begin{array}{c|c} \dfrac{\mathrm{d}}{\mathrm{d}x}\Psi^{(4)}_W(x) = \underline{D}^{(4)}_N(x)\Psi^{(4)}_W(x), & \dfrac{\mathrm{d}}{\mathrm{d}x}\Phi^{(1)}_W(x) = \underline{D}^{(1)}_N(x)\Phi^{(1)}_W(x), \\ \hline \dfrac{\mathrm{d}}{\mathrm{d}x}\Phi^{(4)}_W(x) = D^{(4)}_N(x)\Phi^{(4)}_W(x), & \dfrac{\mathrm{d}}{\mathrm{d}x}\Psi^{(1)}_W(x) = D^{(1)}_N(x)\Psi^{(1)}_W(x). \end{array} \tag{11.6}$$

Under an infinitesimal change of the deformation parameter u_K (the coefficients of the polynomial potential $V(x)$), these vectors satisfy a system of PDEs given by

$$\begin{array}{c|c} \dfrac{\partial}{\partial u_K}\Psi^{(4)}_W(x) = \underline{U}^{N(4)}_K(x)\Psi^{(4)}_W(x), & \dfrac{\partial}{\partial u_K}\Phi^{(1)}_W(x) = \underline{U}^{N(1)}_K(x)\Phi^{(1)}_W(x), \\ \hline \dfrac{\partial}{\partial u_K}\Phi^{(4)}_W(x) = U^{N(4)}_K(x)\Phi^{(4)}_W(x), & \dfrac{\partial}{\partial u_K}\Psi^{(1)}_W(x) = U^{N(1)}_K(x)\Psi^{(1)}_W(x). \end{array} \tag{11.7}$$

Thus the finite subsequence of skew-orthogonal vectors satisfy a system of difference-differential-deformation equation.

The two pairs of matrices $\underline{D}^{(4)}_N(x)$ and $\underline{D}^{(1)}_N(x)$ (similarly $D^{(1)}_N(x)$ and $D^{(4)}_N(x)$), $\underline{U}^{N(4)}_K(x)$ and $\underline{U}^{N(1)}_K(x)$ (similarly $U^{N(1)}_K(x)$ and $U^{N(4)}_K(x)$) and $A^{(4)}_N(x)$ and $A^{(1)}_N(x)$ (similarly $B^{(4)}_N(x)$ and $B^{(1)}_N(x)$) are dual in the sense that they remain invariant under an interchange of $\Psi^{(4)}(x) \mapsto \Phi^{(1)}(x)$ and $\overline{R}^{(4)}(x) \mapsto \overline{R}^{(1)}(y)$ (and similarly for $\Psi^{(1)}(x)$ and $\Phi^{(4)}(x)$).

A look at the system of differential-difference-deformation equations satisfied by the two families of skew-orthogonal functions shows that there exists a duality provided we can establish a mapping between $\Phi^{(4)}(x) \mapsto \Psi^{(1)}(x)$, their derivatives and $\overline{R}^{(4)}(x)/f(y) \mapsto -\overline{R}^{(1)}(y)/f(x)$, where $f(x) = f(y) = 1$ for polynomial weight.

Duality and skew-orthogonal functions. Duality between skew-orthogonal functions was predicted in [36,37] in the context of generalized Christoffel – Darboux formulas and differential systems. In this section, we show the existence of such duality for the quartic and Jacobi family of skew-orthogonal functions.

We expand functions $\phi^{(4)}_m(x)$ and $\psi^{(1)}_m(x)$, skew-orthogonal in the range $[x_1, x_2]$, in a suitable basis of orthogonal polynomials such that:

(i) their derivatives are continuous in the range $[x_1, x_2]$ and vanish at the end-points.

(ii) $\phi_m^{(4)}(x)$ and $\left(\psi_m^{(1)}(x)\right)' \equiv \phi_m^{(1)}(x)$ can be written as $w(x)\pi_m^{(\beta)}(x)$, where $\pi_m^{(\beta)}(x)$ is a polynomial of order m.

(iii) $\phi_m^{(4)}(x)$ and $\psi_m^{(1)}(x)$ are skew-orthonormal in the range $[x_1, x_2]$ with respect to their derivatives.

With these restrictions, we expand Jacobi skew-orthogonal functions $\phi_m^{(4)}(x)$ and $\psi_m^{(1)}(x)$ in terms of Jacobi orthogonal polynomials $P_j^{2a+1,2b+1}(x)$, orthogonal with respect to the weight function $w_{2a+1,2b+1}(x)$ (1.6). For associated Laguerre skew-orthogonal functions, we expand in terms of $L_j^{2a+1}(2x)$, orthogonal with respect to the weight function $w_{2a+1}(x)$ (1.7) while for Gaussian skew-orthogonal functions, the basis chosen is $H_j(x)$ orthogonal with respect to the weight e^{-x^2}. For the quartic weight, we choose polynomials, orthogonal with respect to the weight $\exp[-NV(x)]$, where $V(x)$ is the quartic two-cut potential (1.9). The results obtained show a remarkable similarity between the two families of skew-orthogonal functions corresponding to $\beta = 1$ and 4.

Jacobi skew-orthogonal functions. Jacobi skew-orthogonal functions corresponding to orthogonal and symplectic ensembles are given below. We see that there exists a relation between $\psi_m^{(1)}(x)$ and $\phi_m^{(4)}(x)$ and their derivatives for Jacobi skew-orthogonal functions corresponding to the orthogonal and symplectic ensembles.

For $\beta = 1$, with $(g_m^{(1)})^{1/2}\phi_m^{(1)}(x) = w_{a,b}(x)\pi_m^{(1)}(x)$, we have

$$(g_{2m}^{(1)})^{1/2}\psi_{2m+1}^{(1)}(x) = w_{a+1,b+1}(x)P_{2m}^{2a+1,2b+1}(x), \tag{11.8}$$

$$(g_{2m}^{(1)})^{1/2}\phi_{2m+1}^{(1)}(x) = w_{a,b}(x)[A_{2m+1}P_{2m+1}^{2a+1,2b+1}(x) - B_{2m-1}P_{2m-1}^{2a+1,2b+1}(x)], \tag{11.9}$$

$$(g_{2m}^{(1)})^{1/2}\psi_{2m}^{(1)}(x) = \frac{w_{a+1,b+1}(x)}{A_{2m}}P_{2m-1}^{2a+1,2b+1}(x) + \gamma_{2m-2}^{(2m)}(g_{2m-2}^{(1)})^{1/2}\psi_{2m-2}^{(1)}(x), \quad m \neq 0, \tag{11.10}$$

$$(g_0^{(1)})^{1/2}\psi_0^{(1)}(x) = \int \epsilon(x-y)w_{a,b}(y)\,\mathrm{d}y, \tag{11.11}$$

$$(g_{2m}^{(1)})^{1/2}\phi_{2m}^{(1)}(x) = w_{a,b}(x)P_{2m}^{2a+1,2b+1}(x), \tag{11.12}$$

with

$$g_{2m}^{(1)} = g_{2m+1}^{(1)} = h_{2m}^{2a+1,2b+1}, \qquad m = 0, 1, \ldots. \tag{11.13}$$

For $\beta = 4$, with $(g_m^{(4)})^{1/2}\phi_m^{(4)}(x) = w_{a,b}(x)\pi_m^{(4)}(x)$, we have

$$(g_{2m}^{(4)})^{1/2}\phi_{2m+1}^{(4)}(x) = w_{a+1,b+1}(x)P_{2m-1}^{2a+1,2b+1}(x), \tag{11.14}$$

$$(g_{2m}^{(4)})^{1/2}\psi_{2m+1}^{(4)}(x) = w_{a,b}(x)[A_{2m}P_{2m}^{2a+1,2b+1}(x) - B_{2m-2}P_{2m-2}^{2a+1,2b+1}(x)], \tag{11.15}$$

$$(g_{2m}^{(4)})^{1/2}\phi_{2m}^{(4)}(x) = -\frac{w_{a+1,b+1}(x)}{A_{2m-1}}P_{2m-2}^{2a+1,2b+1}(x) + \gamma_{2m-3}^{(2m-1)}(g_{2m-2}^{(4)})^{1/2}\phi_{2m-2}^{(4)}(x), \tag{11.16}$$

$$(g_{2m}^{(4)})^{1/2}\psi_{2m}^{(4)}(x) = -w_{a,b}(x)P_{2m-1}^{2a+1,2b+1}(x), \tag{11.17}$$

with

$$g_{2m}^{(4)} = g_{2m+1}^{(4)} = h_{2m-1}^{2a+1,2b+1}, \qquad m = 1, 2, \ldots. \tag{11.18}$$

$\gamma_{j-2}^{(j)} \equiv \gamma_j$ and A_j, B_j are given in (4.29) and (4.17) respectively.

$\phi_0^{(4)}(x)$ and $\phi_1^{(4)}(x)$ can be calculated using skew-orthonormality and the Gram–Schmidt method for skew-orthogonal functions. Here we note that skew-orthogonal functions for $\beta = 4$ are lower than those for $\beta = 1$ by order 1. We also observe a minus sign in (11.17) which comes from the anti-symmetric skew-orthonormality condition.

Associated Laguerre skew-orthogonal functions. Associated Laguerre ensembles of random matrices can and do play a significant role in describing real physical systems since they have positive eigenvalues. There exists a simple duality relation between the skew-orthogonal functions $\psi_m^{(1)}(x)$ and $\phi_m^{(4)}(x)$ and their derivatives. Written in terms of the semi-infinite matrices $\Phi^{(\beta)}(x)$ (3.16) and $\Psi^{(\beta)}(x)$ (3.17), we have

$$\Psi^{(1)}(x) = -\sigma_3 \Phi^{(4)}(x), \qquad \Phi^{(1)}(x) = -\sigma_3 \Psi^{(4)}(x), \tag{11.19}$$

where

$$\sigma_3 = \begin{pmatrix} 1 & 0 \\ 0 & -1 \end{pmatrix} \dotplus \cdots \dotplus \tag{11.20}$$

is the semi-infinite Pauli matrix.

We now present the results for the skew-orthogonal functions corresponding to associated Laguerre weights. For $\beta = 1$, with $(g_m^{(1)})^{1/2}\phi_m^{(1)}(x) = w_a(x)\pi_m^{(1)}(x)$, we have

$$(g_{2m+1}^{(1)})^{1/2}\psi_{2m+1}^{(1)}(x) = 2^{a+3/2} w_{a+1}(x) L_{2m}^{2a+1}(2x), \tag{11.21}$$

$$(g_{2m+1}^{(1)})^{1/2}\phi_{2m+1}^{(1)}(x) = 2^{a+1/2} w_a(x)[A_{2m+1}^{\mathrm{L}} L_{2m+1}^{2a+1}(2x) - B_{2m-1}^{\mathrm{L}} L_{2m-1}^{2a+1}(2x)], \tag{11.22}$$

$$(g_{2m}^{(1)})^{1/2}\psi_{2m}^{(1)}(x) = 2^{a+3/2} \frac{w_{a+1}(x)}{A_{2m}^{\mathrm{L}}} L_{2m-1}^{2a+1}(2x) + \gamma_{2m-2}^{(2m)} (g_{2m-2}^{(1)})^{1/2} \psi_{2m-2}^{(1)}(x), \quad m \neq 0, \tag{11.23}$$

$$(g_0^{(1)})^{1/2}\psi_0^{(1)}(x) = 2^{a+1/2} \int_0^\infty \epsilon(x-y) w_a(y)\, \mathrm{d}y, \tag{11.24}$$

$$(g_{2m}^{(1)})^{1/2}\phi_{2m}^{(1)}(x) = 2^{a+1/2} w_a(x) L_{2m}^{2a+1}(2x), \tag{11.25}$$

$$g_{2m}^{(1)} = g_{2m+1}^{(1)} = h_{2m}^{2a+1}, \qquad m = 0, 1, \ldots. \tag{11.26}$$

For $\beta = 4$, with $(g_m^{(4)})^{1/2}\phi_m^{(4)}(x) = w_a(x)\pi_m^{(4)}(x)$, we have

$$(g_{2m+1}^{(4)})^{1/2}\phi_{2m+1}^{(4)}(x) = 2^{a+3/2} w_{a+1}(x) L_{2m}^{2a+1}(2x), \tag{11.27}$$

$$(g_{2m+1}^{(4)})^{1/2}\psi_{2m+1}^{(4)}(x) = 2^{a+1/2} w_a(x)[A_{2m+1}^{\mathrm{L}} L_{2m+1}^{2a+1}(2x) - B_{2m-1}^{\mathrm{L}} L_{2m-1}^{2a+1}(2x)], \tag{11.28}$$

$$(g_{2m}^{(4)})^{1/2}\phi_{2m}^{(4)}(x) = -2^{a+3/2} \frac{w_{a+1}(x)}{A_{2m}^{\mathrm{L}}} L_{2m-1}^{2a+1}(2x) + \gamma_{2m-3}^{(2m-1)} (g_{2m-2}^{(4)})^{1/2} \phi_{2m-2}^{(4)}(x), \quad m \neq 0, \tag{11.29}$$

$$(g_0^{(4)})^{1/2}\phi_0^{(4)}(x) = -2^{a+1/2} \int_0^\infty \epsilon(x-y) w_a(y)\, \mathrm{d}y, \tag{11.30}$$

$$(g_{2m}^{(4)})^{1/2}\psi_{2m}^{(4)}(x) = -2^{a+1/2}w_a(x)L_{2m}^{2a+1}(2x), \tag{11.31}$$

$$g_{2m}^{(4)} = g_{2m+1}^{(4)} = h_{2m}^{2a+1}, \qquad m = 0, 1, \dots. \tag{11.32}$$

Here $\gamma_{j-2}^{(j)} \equiv \gamma_j^{\mathrm{L}}$, A_j^{L} and B_j^{L} are given in (4.46), (4.47) and (4.48) respectively.

Gaussian skew-orthogonal functions. For Gaussian weight, with $w(x) = e^{-x^2/2}$ and $\beta = 1$, the skew-orthogonal functions can be expressed in terms of orthogonal Hermite polynomials as:

$$(g_{2m+1}^{(1)})^{1/2}\psi_{2m+1}^{(1)}(x) = w(x)H_{2m}(x), \tag{11.33}$$

$$(g_{2m+1}^{(1)})^{1/2}\phi_{2m+1}^{(1)}(x) = w(x)[-(1/2)H_{2m+1}(x) + 2mH_{2m-1}(x)], \tag{11.34}$$

$$\begin{aligned}(g_{2m}^{(1)})^{1/2}\psi_{2m}^{(1)}(x) &= -2w(x)H_{2m-1}(x)\\ &\quad + 2(2m-1)(g_{2m-2}^{(1)})^{1/2}\psi_{2m-2}^{(1)}(x), \quad m \neq 0,\end{aligned} \tag{11.35}$$

$$(g_0^{(1)})^{1/2}\psi_0^{(1)}(x) = \int_{-\infty}^{\infty} \epsilon(x-y)w(y)\, \mathrm{d}y, \tag{11.36}$$

$$(g_{2m}^{(1)})^{1/2}\phi_{2m}^{(1)}(x) = w(x)H_{2m}(x), \tag{11.37}$$

$$g_{2m}^{(1)} = g_{2m+1}^{(1)} = h_{2m}, \qquad m = 0, 1, \dots. \tag{11.38}$$

For $\beta = 4$,

$$(g_{2m+1}^{(4)})^{1/2}\phi_{2m+1}^{(4)}(x) = w(x)H_{2m+1}(x), \tag{11.39}$$

$$(g_{2m+1}^{(4)})^{1/2}\psi_{2m+1}^{(4)}(x) = w(x)[-(1/2)H_{2m+2}(x) + (2m+1)H_{2m}(x)], \tag{11.40}$$

$$(g_{2m}^{(4)})^{1/2}\phi_{2m}^{(4)}(x) = 2w(x)H_{2m}(x) + 4m(g_{2m-2}^{(4)})^{1/2}\phi_{2m-2}^{(4)}(x), \tag{11.41}$$

$$(g_{2m}^{(4)})^{1/2}\psi_{2m}^{(4)}(x) = -w(x)H_{2m+1}(x), \tag{11.42}$$

$$g_{2m}^{(4)} = g_{2m+1}^{(4)} = h_{2m+1}, \qquad m = 0, 1, \dots. \tag{11.43}$$

Thus for the Gaussian weight, skew-orthogonal functions for $\beta = 4$ are higher than those for $\beta = 1$ by order 1. This is exactly the opposite of what we saw for Jacobi skew-orthogonal functions.

Quartic weight. For $w(x) = \exp[-NV(x)]$, where the quartic two-cut potential $V(x)$ is defined by (1.9), the skew-orthogonal functions can be expressed in terms of the corresponding orthogonal polynomials $\phi_n^{(2)}(x) = w(x)P_n^{(2)}(x)$. Here, we use the formula (for $d = 2$):

$$\frac{\mathrm{d}}{\mathrm{d}x}\phi_n^{(2)}(x) = \sum_{m=n-2d+1}^{n+2d-1} P_{n,m}^{(2)}\phi_m^{(2)}(x), \qquad P^{(2)} = -P^{(2)\mathrm{t}}. \tag{11.44}$$

We get

$$\psi_{2m+1}^{(1)}(x) = \frac{\sqrt{R_{2m-1}}}{x}\left[\phi_{2m-1}^{(2)}(x) - \frac{x\psi_1^{(1)}(x)}{\phi_{2m-2}^{(2)}(0)}\right], \tag{11.45}$$

$$\phi_{2m}^{(1)}(x) = \frac{x\phi_{2m-1}^{(2)}(x)}{\sqrt{R_{2m-1}}}, \qquad m > 1, \tag{11.46}$$

$$g_{2m}^{(1)} = h_{2m-1}. \tag{11.47}$$

$\phi_{2m+1}^{(1)}(x)$ and $\psi_{2m}^{(1)}(x)$ can be obtained by taking derivative of $\psi_{2m+1}^{(1)}(x)$ and using skew-orthonormality respectively. $\phi_j^{(1)}(x)$ and $\psi_j^{(1)}(x)$ for $j \leq 3$ can be calculated using the Gram – Schmidt technique. However, the condition $(\psi_3^{(1)}(x), \phi_0^{(1)}(x)) = 0$ gives an additional term in (11.45).

For $\beta = 4$, we have

$$\phi_{2m+1}^{(4)}(x) = \frac{\sqrt{R_{2m+2}}}{x}\left[\phi_{2m+2}^{(2)}(x) - \phi_{2m+2}^{(2)}(0)\exp[-NV(x)]\right], \tag{11.48}$$

$$\psi_{2m}^{(4)}(x) = -\frac{x\phi_{2m+2}^{(2)}(x)}{\sqrt{R_{2m+2}}}, \tag{11.49}$$

$$g_{2m}^{(4)} = h_{2m+2}. \tag{11.50}$$

$\phi_{2m}^{(4)}(x)$ and $\psi_{2m+1}^{(4)}(x)$ can be derived by integrating and differentiating $\psi_{2m}^{(4)}(x)$ and $\phi_{2m+1}^{(4)}(x)$ respectively.

Conjecture. This prompts us to write the following conjecture for general d:

$$\psi_{2m+1}^{(1)}(x) \simeq \sqrt{R_{2m-2d+3}\cdots R_{2m-d+1}}\left[\frac{\phi_{2m-d+1}^{(2)}(x)}{x^{d-1}}\right], \tag{11.51}$$

$$\phi_{2m}^{(1)}(x) = \frac{x^{d-1}\phi_{2m-d+1}^{(2)}(x)}{\sqrt{R_{2m-2d+3}\cdots R_{2m-d+1}}}, \qquad m > d-1, \tag{11.52}$$

$$g_{2m}^{(1)} = h_{2m-d+1}. \tag{11.53}$$

$\phi_{2m+1}^{(1)}(x)$ and $\psi_{2m}^{(1)}(x)$ can be obtained by taking derivative of $\psi_{2m+1}^{(1)}(x)$ and using skew-orthonormality respectively. Here, we have written $Q_{2m-2d+2,2m-2d+3}^{(2)} \equiv \sqrt{R_{2m-2d+3}}$, where $Q^{(2)}$ is the symmetric tridiagonal Jacobi matrix.

For $\beta = 4$,

$$\phi_{2m+1}^{(4)}(x) = \frac{\sqrt{R_{2m+2}\cdots R_{2m+d}}}{x^{d-1}\sqrt{2}}\left[\phi_{2m+d}^{(2)}(x) - \phi_{2m+d}^{(2)}(0)\exp[-NV(x)]\right], \tag{11.54}$$

$$\psi_{2m}^{(4)}(x) = -\frac{x^{d-1}\phi_{2m+d}^{(2)}(x)}{\sqrt{2R_{2m+2}\cdots R_{2m+d}}}, \tag{11.55}$$

$$g_{2m}^{(4)} = h_{2m+d}. \tag{11.56}$$

$\phi_{2m}^{(4)}(x)$ and $\psi_{2m+1}^{(4)}(x)$ can be derived by integrating and differentiating $\psi_{2m}^{(4)}(x)$ and $\phi_{2m+1}^{(4)}(x)$ respectively.

CHAPTER 12

Conclusion

In this book, we develop the theory of skew-orthogonal polynomials and use them to study non-Gaussian ensembles of random matrices. We observe that skew-orthogonal functions satisfy $(2d + 1)$-term recursion relations in the quaternion space (as compared to three-term recursion relations in ordinary space for orthogonal polynomials), for weight functions defined in (3.23) – (3.25). We use this to obtain an analog of the Christoffel – Darboux formula for skew-orthogonal functions, containing $2d - 1$ quaternions (as compared to one term for orthogonal functions). We call it the generalized Christoffel – Darboux formula.

We also show that a finite subsequence of skew-orthogonal functions with polynomial weight satisfy a mutually compatible system of differential-difference-deformation equation and admit a common fundamental solution. This points out its intimate connection with integrable systems and Riemann – Hilbert (monodromy) problems.

However, the usefulness of the mutually compatible system of differential-difference-deformation equation in the statistical analysis of random matrix theory is still at a "theoretical" (read nonimplementable) level, since the asymptotics for the skew-orthogonal functions are still not computable via the Riemann – Hilbert approach.

Matrix integral representations of skew-orthogonal polynomials also appear to be promising for the asymptotic analysis of skew-orthogonal functions. In fact, Eynard [30] used the method of resolvents to obtain asymptotics for the one-cut polynomial potential. Purists call it nonrigorous.

To circumvent the problem of finding the asymptotics of skew-orthogonal functions, we obtain explicit mapping relations between skew-orthogonal and orthogonal functions. This helps us to obtain rigorously Plancherel – Rotach type asymptotic formulas for skew-orthogonal functions and their recursion coefficients using well-known results on orthogonal polynomials. As a by-product, we demonstrate duality relations between the two families of skew-orthogonal functions.

For statistical analysis of random matrix ensembles, we use the generalized Christoffel – Darboux formula and the asymptotic results of skew-orthogonal functions to demonstrate universality in the bulk of the spectrum. Universality at the edge is still not done using this technique. The level densities for these random matrix ensembles has been calculated by both the polynomial and nonpolynomial (resolvent) methods.

Here, we would like to mention that Deift and Gioev have proved "universality" [20, 21] for the orthogonal and symplectic ensembles using the Sener – Verbaarschot – Widom formalism [72, 78]. The method involves expressing the kernel functions for these two ensembles in terms of that of the unitary ensemble. Asymptotic properties of orthogonal polynomials [18] are then used. The method is rigorous but cumbersome.

To conclude, this work makes a significant contribution to our understanding of some of the fundamental properties of skew-orthogonal functions. But a lot more remains to be done. In this regard, we recall that Dyson, while introducing quaternions and skew-orthogonal polynomials [27] to generalize the work of Fox and Kahn [32] to $\beta = 1$ and 4, stated "in order to make the explicit formulas of this section useful, we need to develop the theory of skew-orthogonal polynomials until it becomes a working tool as handy as the existing theory of orthogonal polynomials." With the discovery of the recursion relations, the generalized Christoffel–Darboux formula and the asymptotics of skew-orthogonal polynomials, it would perhaps not be out of context to claim that skew-orthogonal polynomials have indeed become a "handy tool." Judgement on the "as handy as orthogonal polynomials" part is best left to the readers' discretion.

APPENDIX A

Proofs of (5.7), (5.12), and (5.19)

PROOF OF (5.7). The asymptotic form of the associated Laguerre polynomial is given by (5.6). For convenience, we call the amplitude part $A(\theta, m)$ and the argument part $f_m(\theta) + \alpha$. Thus (5.6) can be rewritten as

$$\phi_m^{(2)}(x) \equiv (h_m^{(a)})^{-1/2}\big(w_a(x)\big)^{1/2} L_m^{(a)}(x) = A(\theta, m)\sin[f_m(\theta_m) + \alpha], \tag{A.1}$$

where

$$A(\theta, m) = \frac{(-1)^m}{\sqrt{2\pi \sin\theta_m \cos\theta_m}},$$
$$f_m(\theta_m) = \left(m + \frac{a+1}{2}\right)(\sin 2\theta_m - 2\theta_m),$$
$$\alpha = \frac{3\pi}{4},$$

where $x = (4m + 2a + 2)\cos^2\theta$. Here, x does not depend on m but θ does. Hence differentiating with respect to θ and putting $\Delta m = \pm 1$, we have

$$\Delta\theta_m = \pm\frac{1}{2}\left[\left(m + \frac{a+1}{2}\right)\tan\theta_m\right]^{-1} \simeq \pm\frac{1}{2m\tan\theta_m} \tag{A.2}$$

We also get

$$\left(\frac{\partial f_m(\theta_m)}{\partial m}\right) = (\sin 2\theta_m - 2\theta_m) \tag{A.3}$$

and

$$\left(\frac{\partial f_m(\theta_m)}{\partial \theta_m}\right) = -4\left(m + \frac{a+1}{2}\right)\sin^2\theta_m. \tag{A.4}$$

Using (A.1)–(A.4), we can write

$$\begin{aligned}\phi_{m\pm1}^{(2)}(x) &= A(\theta_m, m)\sin[f_{m\pm1}(\theta_{m\pm1}) + \alpha] \\ &= A(\theta_m, m)\sin\left[f_m(\theta_m) \pm \frac{\partial f_m(\theta_m)}{\partial m}\Delta m \pm \frac{\partial f_m(\theta_m)}{\partial \theta_m}\Delta\theta_m + \alpha\right] \\ &= A(\theta_m, m)\sin\left[\left(m + \frac{a+1}{2}\right)(\sin 2\theta_m - 2\theta_m)\right. \\ &\qquad\qquad \left. \pm (\sin 2\theta_m - 2\theta_m) \mp \sin 2\theta_m + \frac{3\pi}{4}\right].\end{aligned} \tag{A.5}$$

Thus we have

$$\phi_{m\pm1}^{(2)}(x) = \frac{(-1)^m}{\sqrt{2\pi\sin\theta_m\cos\theta_m}}\sin\left[\left(m + \frac{a+1}{2}\right)(\sin 2\theta_m - 2\theta_m) \mp 2\theta_m + \frac{3\pi}{4}\right]. \quad \square \tag{A.6}$$

PROOF OF (5.12). The asymptotic form of the Hermite polynomial is given by (5.11). For convenience, we call the amplitude part $A(\theta, m)$ and the argument part $f_m(\theta) + \alpha$. Thus (5.11) can be rewritten as

$$\phi_m^{(2)}(x) \equiv (h_m^{(a)})^{-1/2}\big(w_G(x)\big)^{1/2} H_m(x) = A(\theta, m)\sin[f_m(\theta_m) + \alpha], \tag{A.7}$$

where

$$A(\theta, m) = \frac{1}{\sqrt{\pi \sin\theta}}\left(\frac{2}{m}\right)^{1/4},$$
$$f_m(\theta_m) = \left(\frac{m}{2} + \frac{1}{4}\right)(\sin 2\theta_m - 2\theta_m),$$
$$\alpha = \frac{3\pi}{4}$$

where $x = \sqrt{(2m+1)}\cos\theta$. Here, x does not depend on m but θ does. Hence differentiating with respect to θ and putting $\Delta m = \pm 1$, we have

$$\Delta\theta_m = \pm\frac{1}{(2m+1)\tan\theta_m} \simeq \pm\frac{1}{2m\tan\theta_m}. \tag{A.8}$$

We also get

$$\left(\frac{\partial f_m(\theta_m)}{\partial m}\right) = \frac{1}{2}(\sin 2\theta_m - 2\theta_m) \tag{A.9}$$

and

$$\left(\frac{\partial f_m(\theta_m)}{\partial \theta_m}\right) = -4\left(\frac{m}{2} + \frac{1}{4}\right)\sin^2\theta_m. \tag{A.10}$$

Using (A.7) – (A.10), we can write

$$\begin{aligned}
\phi_{m\pm1}^{(2)}(x) &= A(\theta_m, m)\sin[f_{m\pm1}(\theta_{m\pm1}) + \alpha] \\
&= A(\theta_m, m)\sin\left[f_m(\theta_m) \pm \frac{\partial f_m(\theta_m)}{\partial m}\Delta m \pm \frac{\partial f_m(\theta_m)}{\partial\theta_m}\Delta\theta_m + \alpha\right] \\
&= A(\theta_m, m)\sin\left[\left(\frac{m}{2} + \frac{1}{4}\right)(\sin 2\theta_m - 2\theta_m)\right. \\
&\qquad\left. \pm\frac{1}{2}(\sin 2\theta_m - 2\theta_m) \mp \frac{\sin 2\theta_m}{2} + \frac{3\pi}{4}\right].
\end{aligned} \tag{A.11}$$

Thus we have

$$\begin{aligned}
&\phi_{m\pm1}^{(2)}(x) \\
&= \frac{1}{\sqrt{\pi\sin\theta_m}}\left(\frac{2}{m}\right)^{1/4}\sin\left[\left(\frac{j}{2} + \frac{1}{4}\right)(\sin 2\theta_m - 2\theta_m) \mp \theta_m + \frac{3\pi}{4}\right]. \quad \square
\end{aligned} \tag{A.12}$$

PROOF OF (5.19). The asymptotic form of orthogonal functions corresponding to the quartic weight is given by (5.16). For convenience, we call the amplitude part $A(x, \theta_m)$ and the argument part $f_m(\theta)$.

$$\phi_m^{(2)}(x) = A(x, \theta_m)\big[\cos\big(f_m(\theta)\big) + O(N^{-1})\big], \tag{A.13}$$

where

$$f_m(\theta) = \left(\frac{m + \frac{1}{2}}{2}\right)\left(\frac{\sin(2\theta)}{2} - \theta\right) - (-1)^m\frac{\chi}{4} + \frac{\pi}{4}.$$

We know that

$$2\sqrt{\lambda' g}\cos\theta = gx^2 + t, \qquad \lambda' = \frac{m+\frac{1}{2}}{N}, \qquad \lambda \equiv \frac{m}{N}, \tag{A.14}$$

where, for a given x, θ varies with λ. Hence differentiating with respect to θ and putting $N\Delta\lambda' = \pm 1$, we have

$$\Delta\theta_m = \pm\frac{1}{2\lambda' \tan\theta_m}. \tag{A.15}$$

We also get

$$\left(\frac{\partial f_m(\theta_m)}{\partial \lambda'}\right) = \frac{N}{2}\left(\frac{\sin 2\theta_m}{2} - \theta_m\right) \tag{A.16}$$

and

$$\left(\frac{\partial f_m(\theta_m)}{\partial \theta_m}\right) = -N\lambda' \sin^2\theta_m. \tag{A.17}$$

Using (A.15)–(A.17), we can write

$$\begin{aligned} \phi^{(2)}_{m\pm1}(x) &= A(x,\theta_m)\cos[f_{m\pm1}(\theta_{m\pm1})] \\ &= A(x,\theta_m)\cos\left[f_m(\theta_m) \pm \frac{\partial f_m(\theta_m)}{\partial\lambda'}\Delta\lambda' \pm \frac{\partial f_m(\theta_m)}{\partial\theta_m}\Delta\theta_m\right] \\ &= A(x,\theta_m)\cos\left[\frac{(m+\frac{1}{2})}{2}\left(\frac{\sin(2\theta_m)}{2} - \theta_m\right)\right. \\ &\qquad \left.\pm\frac{1}{2}\left(\frac{\sin(2\theta_m)}{2} - \theta_m\right) \mp \frac{1}{4}\sin 2\theta_m - (-1)^m\frac{\chi}{4} + \frac{\pi}{4}\right]. \end{aligned} \tag{A.18}$$

Thus we have

$$\begin{aligned} &\phi^{(2)}_{m\pm1}(x) \\ &= \frac{2C_m\sqrt{x}}{\sqrt{\sin\theta_m}}\cos\left[\frac{(m+\frac{1}{2})}{2}\left(\frac{\sin 2\theta_m}{2} - \theta_m\right) \mp \frac{\theta_m}{2} - (-1)^m\frac{\chi}{4} + \frac{\pi}{4}\right]. \end{aligned} \tag{A.19}$$

Here we note that for a given x, χ remains constant under the variation of θ and m. □

APPENDIX B

Associated Laguerre and Gaussian Results as Limiting Cases of Jacobi Skew-orthogonal Polynomials

In this appendix we prove the Jacobi results (8.6) – (8.15) and the associated Laguerre and Gaussian results for the skew-orthogonal polynomials given in Chapters 6, 8 for $\beta = 1, 4$ respectively by taking the limit.

For associated Laguerre and Hermite weight functions one can directly follow the above procedure, or more simply take the limits of the Jacobi results as discussed below. For associated Laguerre, note first that

$$w_a(x) = \lim_{b\to\infty} 2^{-a-b} b^a w_{ab}(1 - 2b^{-1}x), \tag{B.1}$$

$$L_j^{(a)}(x) = \lim_{b\to\infty} P_j^{a,b}(1 - 2b^{-1}x), \tag{B.2}$$

$$k_j^{(a)} = \lim_{b\to\infty} \left(-\frac{2}{b}\right)^j k_j^{a,b}, \tag{B.3}$$

$$h_j^{(a)} = \lim_{b\to\infty} \frac{b^{a+1}}{2^{a+b+1}} h_j^{a,b}. \tag{B.4}$$

Thus for skew-orthogonal functions we have (in terms of the Jacobi skew-orthogonal functions $\phi_j^{a,b}$, $\psi_j^{a,b}$),

$$\phi_j^{(1)}(x) = \lim_{b\to\infty} (-1)^j 2^{-b+1/2} b^a \phi_j^{a,b}(1 - 2b^{-1}x), \tag{B.5}$$

$$\psi_j^{(1)}(x) = \lim_{b\to\infty} (-1)^{j-1} 2^{-b-1/2} b^{a+1} \psi_j^{a,b}(1 - 2b^{-1}x), \tag{B.6}$$

giving thereby (4.40) – (4.50). Similarly, for the Hermite weight, note that (with $j = 2m, 2m+1$)

$$e^{-x^2/2} = \lim_{a\to\infty} w_{a,a}\left(\frac{x}{\sqrt{2a}}\right), \tag{B.7}$$

$$H_j(x) = \lim_{a\to\infty} 2^j j! a^{-j/2} P_j^{a,a}\left(\frac{x}{\sqrt{a}}\right), \tag{B.8}$$

$$k_j = \lim_{a\to\infty} 2^j j! a^{-j} k_j^{a,a}, \tag{B.9}$$

$$h_j = \lim_{a\to\infty} (2^j j!)^2 a^{-j+1/2} h_j^{a,a}, \tag{B.10}$$

$$\phi_j^{(1)}(x) = \lim_{a\to\infty} 2^{2m}(2m)!(2a)^{-j/2} \phi_j^{a,a}\left(\frac{x}{\sqrt{2a}}\right), \tag{B.11}$$

$$\psi_j^{(1)}(x) = \lim_{a\to\infty} 2^{2m}(2m)!(2a)^{-(j-1)/2} \psi_j^{a,a}\left(\frac{x}{\sqrt{2a}}\right), \tag{B.12}$$

giving the Hermite results (4.52) – (4.57).

For $\beta = 4$ with the Jacobi weight function, we expand $\pi_j^{(4)\prime}(x)$ as

$$\pi_j^{(4)\prime}(x) = P_{j-1}^{a,b}(x) + \sum_{k=0}^{j-1} \eta_k^{(j)} \pi_k^{(4)\prime}(x), \tag{B.13}$$

so that (8.10) gives

(B.14)

$$\pi_j^{(4)}(x) = \frac{2}{j+a+b-1}\left[D_j P_j^{a,b}(x) + E_j P_{j-1}^{a,b}(x) + F_j P_{j-2}^{a,b}(x)\right] + \sum_{k=0}^{j-1} \eta_k^{(j)} \pi_k^{(4)}(x).$$

Then the orthogonality of the $P_j^{a,b}(x)$ and the skew-orthogonality of the $\pi_j^{(4)}(x)$ give

$$\eta_k^{(2m)} = 0, \tag{B.15}$$

for $k \neq 2m-2$, and

$$\eta_k^{(2m+1)} = 0 \tag{B.16}$$

for $k \neq 2m$. Also $\eta_{2m}^{(2m+1)}$, being arbitrary, is chosen to be zero. Thus $\eta_{2m-2}^{(2m)} \equiv \eta_{2m}$ is the only nonzero coefficient, giving thereby (8.6) – (8.9). The skew-orthogonality of $\pi_{2m}^{(4)}(x)$ and $\pi_{2m-1}^{(4)}(x)$ gives

$$\eta_{2m} = \frac{1}{g_{2m-2}}\left[\frac{2D_{2m-1}}{2m+a+b-2} h_{2m-1}^{a,b} - \frac{2F_{2m}}{2m+a+b-1} h_{2m-2}^{a,b}\right], \tag{B.17}$$

while the normalization is given by

$$g_{2m}^{(4)} = \left[\frac{2D_{2m}}{2m+a+b-1} h_{2m}^{a,b} - \frac{2F_{2m+1}}{2m+a+b} h_{2m-1}^{a,b}\right], \tag{B.18}$$

confirming thereby (8.14), (8.15). To prove the Jacobi result (8.10), we note that the first step is given in a differential form in [77], while for the second step we use [1]

$$(2j+a+b)P_j^{a,b-1}(x) = (j+a+b)P_j^{a,b}(x) + (j+a)P_{j-1}^{a,b}(x), \tag{B.19}$$

$$(2j+a+b)P_j^{a-1,b}(x) = (j+a+b)P_j^{a,b}(x) - (j+b)P_{j-1}^{a,b}(x). \tag{B.20}$$

This completes the proof of (8.6) – (8.15).

The associated Laguerre results (8.37) – (8.42) derive directly by using the limits (B.1) – (B.4) in (8.6) – (8.15), while the Hermite results (8.64) – (8.67) derive from the limits

$$\mathrm{e}^{-2x^2} = \lim_{a\to\infty} w_{a,a}(x\sqrt{2/a}), \tag{B.21}$$

$$H_j(x\sqrt{2}) = \lim_{a\to\infty} 2^j j! a^{-j/2} P_j^{a,a}(x\sqrt{2/a}). \tag{B.22}$$

APPENDIX C

Proofs of (10.2) – (10.9)

In this appendix we outline a proof of the matrix-integral representations (10.2) – (10.9) of the polynomials. The Vandermonde determinant and its fourth power can be written as

$$\Delta(x_1, \dots, x_N) = \det[x_\mu^{N-\nu}]_{\mu,\nu=1,\dots,N}, \tag{C.1}$$

$$(\Delta(x_1, \dots, x_N))^4 = \det[x_\mu^{2N-\nu}, (2N-\nu)x_\mu^{2N-\nu-1}]_{\mu=1,\dots,N,\nu=1,\dots,2N}. \tag{C.2}$$

For $\beta = 2$, (10.2) represents orthogonal polynomials with the weight $w(x)$ if

$$\int x^k P_j(x) w(x) \, \mathrm{d}x = 0, \qquad k = 0, 1, \dots, j-1. \tag{C.3}$$

Using the joint-probability distribution (1.12) result, in the definition of average (10.1) and using the determinant in (10.2), the integral in (C.3) is proportional to

$$\begin{aligned} &\int \mathrm{d}x_1 \cdots \int \mathrm{d}x_{j+1} \, (x_{j+1})^k \Delta(x_1, \dots, x_j) \Delta(x_1, \dots, x_{j+1}) \prod_{\mu=1}^{j+1} w(x_\mu) \\ &= \frac{1}{(j+1)!} \int \mathrm{d}x_1 \cdots \int \mathrm{d}x_{j+1} \left(\sum_P \epsilon_P (x_{i_{j+1}})^k \Delta(x_{i_1}, \dots, x_{i_j}) \right) \\ &\qquad \times \Delta(x_1, \dots, x_{j+1}) \prod_{\mu=1}^{j+1} w(x_\mu), \end{aligned} \tag{C.4}$$

where $\sum_P$ is summation over all permutations $(x_{i_1}, \dots, x_{i_{j+1}})$ of $(x_1, \dots, x_{j+1})$ and ϵ_P $(= \pm 1)$ is the sign of the permutation, equal to the change of sign in $\Delta(x_1, \dots, x_{j+1})$ after the permutation. The summation term in (C.4) can be written as

$$\sum_P \epsilon_P (x_{i_{j+1}})^k \Delta(x_{i_1}, \dots, x_{i_j}) = (-1)^j \det \begin{pmatrix} x_1^k & x_2^k & \dots & x_{j+1}^k \\ x_1^{j-1} & x_2^{j-1} & \dots & x_{j+1}^{j-1} \\ \vdots & \vdots & \ddots & \vdots \\ 1 & 1 & \dots & 1 \end{pmatrix}, \tag{C.5}$$

which is zero for $k = 0, 1, \dots, j-1$, thereby proving (C.3) and hence (10.2).

For $\beta = 1$, we consider (10.3), (10.4) for the even-N case; a similar consideration would apply to (10.5) – (10.7) for the odd-N case. The $\pi_j^{(1)}(x)$ of (10.3), (10.4) represent skew-orthogonal polynomials of the $\beta = 1$ type with the weight $w(x)$ if

$$\iint \mathrm{d}x \, \mathrm{d}y \, \epsilon(x-y) y^k \pi_j^{(1)}(x) w(x) w(y) = 0, \tag{C.6}$$

for $k = 0, \dots, 2m-1$ and also for $k = j$ for both $j = 2m, 2m+1$. The integrals in (10.3), (10.4) involve $|\Delta(x_1, \dots, x_{2m})|$ and therefore Mehta's method of integration over alternate variables [53] can be used. For $j = 2m$, the integral in (C.6) is proportional to

(C.7)
$$
\begin{aligned}
&\int dx_1 \cdots dx_{2m+2}\, \epsilon(x_{2m+1} - x_{2m+2})(x_{2m+2})^k \\
&\qquad \times \left(\prod_{\nu=1}^{2m}(x_{2m+1} - x_\nu)\right)|\Delta(x_1, \dots, x_{2m})| \prod_{\mu=1}^{2m+2} w(x_\mu) \\
&= (2m)! \int_{x_1 \le x_2 \le \cdots \le x_{2m}} dx_1 \cdots \int dx_{2m+2}\, \epsilon(x_{2m+1} - x_{2m+2})(x_{2m+2})^k \\
&\qquad \times \left(\prod_{\mu=1}^{2m+2} w(x_\mu)\right)\Delta(x_1, \dots, x_{2m+1}) \\
&= \frac{1}{2}\frac{(2m)!}{m!} \int dx_1\, dx_3 \cdots dx_{2m+1} \det\begin{pmatrix} 0 & 0 & \dots x_{2m+1}^k & F_k(x_{2m+1}) \\ x_1^{2m} & F_{2m}(x_1) \dots & x_{2m}^{2m} & F_{2m}(x_{2m+1}) \\ \vdots & \vdots & \ddots\ \vdots & \vdots \\ 1 & F_0(x_1) \dots & 1 & F_0(x_{2m+1}) \end{pmatrix} \\
&\qquad \times \prod_{i=0}^{m} w(x_{2i+1}) \\
&= \frac{1}{2}\frac{(2m)!}{m!(m+1)!} \int dx_1\, dx_3 \cdots dx_{2m+1} \det\begin{pmatrix} x_1^k & F_k(x_1) & \dots \\ x_1^{2m} & F_{2m}(x_1) & \dots \\ x_1^{2m-1} & F_{2m-1}(x_1) & \dots \\ \vdots & \vdots & \vdots \\ 1 & F_0(x_1) & \dots \end{pmatrix} \\
&\qquad \times \prod_{i=0}^{m} w(x_{2i+1}).
\end{aligned}
$$

In the second and third steps the above-mentioned Mehta's method of integration over alternate variables is used, where $F_k(x)$ is given by

$$F_k(x) = \int_x^\infty y^k w(y)\, dy. \tag{C.8}$$

In the last step of (C.7) all permutations of $(x_1, x_3, \dots, x_{2m+1})$ have been used. The determinant in the last step is zero for $k = 0, \dots, 2m$, thereby proving (C.6) and (10.3). For $j = 2m+1$, the first integral in (C.7) has, in the integrand, the extra factor $(x_{2m+1} + \sum x_\nu)$ so that $\Delta(x_1, \dots, x_{2m+1})$ in the second form is replaced by

$$\det\begin{pmatrix} x_1^{2m+1} & \dots & x_{2m+1}^{2m+1} \\ x_1^{2m-1} & \dots & x_{2m+1}^{2m-1} \\ \vdots & \dots & \vdots \\ 1 & \dots & 1 \end{pmatrix} = \left(\sum_{\mu=1}^{2m+1} x_\mu\right)\Delta(x_1, \dots, x_{2m+1}). \tag{C.9}$$

Then the second row in the determinant of the last step (C.7) is replaced by $x_1^{2m+1}, F_{2m+1}(x_1), \dots,$ other rows remaining the same. Again (C.6) for $k = 0, \dots, 2m-1, 2m+1$ and hence (10.4) are verified.

For $\beta = 4$, (10.8) – (10.9) represent the skew-orthogonal polynomials if

$$\int \mathrm{d}x \, \{x^k \pi_j^{(4)\prime}(x) - k x^{k-1} \pi_j^{(4)}(x)\} w(x) = 0, \tag{C.10}$$

for $k = 0, \ldots, 2m-1$ and also for $k = j$ for $j = 2m$, $2m+1$ both. In this case we use (C.2) in the joint-probability density (1.12). For $j = 2m$, the integral in (C.10) is proportional to

$$\begin{aligned}
&\int \mathrm{d}x_1 \cdots \mathrm{d}x_{2m} \, \mathrm{d}x_{2m+1} \\
&\qquad \left\{ x_{2m+1}^k \frac{\mathrm{d}}{\mathrm{d}x_{2m+1}} \prod_{\nu=1}^{m} (x_{2m+1} - x_\nu)^2 - k x_{2m+1}^{k-1} \prod_{\nu=1}^{m} (x_{2m+1} - x_\nu)^2 \right\} \\
&\qquad\qquad \times \left(\prod_{\mu=1}^{2m+1} w(x_\mu) \right) \big(\Delta(x_1, \ldots, x_m) \big)^4 \\
&= \int \mathrm{d}x_1 \cdots \mathrm{d}x_{2m+1} \det \begin{pmatrix} 0 & 0 & \ldots & x_{2m+1}^k & k x_{2m+1}^{k-1} \\ x_1^{2m} & 2m x_1^{2m-1} & \ldots & x_{2m+1}^{2m} & 2m x_{2m+1}^{2m-1} \\ \vdots & \vdots & \ldots & \vdots & \vdots \\ 1 & 0 & \ldots & 1 & 0 \end{pmatrix} \prod_{\mu=1}^{2m+1} w(x_\mu) \\
&= \frac{1}{(m+1)!} \int \mathrm{d}x_1 \cdots \mathrm{d}x_{2m+1} \det \begin{pmatrix} x_1^k & k x_1^{k-1} & \ldots \\ x_1^{2m} & 2m x_1^{2m-1} & \ldots \\ \vdots & \vdots & \vdots \\ 1 & 0 & \ldots \end{pmatrix} \prod_{\mu=1}^{2m+1} w(x_\mu),
\end{aligned} \tag{C.11}$$

where the last step is by a permutation of all the variables in the first step. The determinant in the last step is again zero for $k = 0, \ldots, 2m$, confirming (C.10) and hence (10.8). For $j = 2m+1$, we have the additional term $(x_{2m+1} + 2\sum x_\nu)$ with $\prod (x - x_\nu)^2$. In this case the second rows of both the determinants of (C.11) are replaced by $(x_1^{2m+1}, (2m+1) x_1^{2m}, \ldots)$, the last determinant being then zero for $k = 0, \ldots, 2m-1$, and $2m+1$. Thus (10.9) is verified.

Bibliography

1. M. Abramowitz and I. A. Stegun, *Handbook of mathematical functions with formulas, graphs, and mathematical tables*, Dover, New York, 1965.
2. M. Adler, P. J. Forrester, T. Nagao, and P. van Moerbeke, *Classical skew orthogonal polynomials and random matrices*, J. Statist. Phys. **99** (2000), no. 1-2, 141–170.
3. R. Balian, *Random matrices and information theory*, Nuovo Cimento B (10) **57** (1968), no. 1, 183–193.
4. C. W. J. Beenakker, *Random-matrix theory of quantum transport*, Rev. Modern Phys. **69** (1997), no. 3, 731–808.
5. M. Bergère and B. Eynard, *Mixed correlation function and spectral curve for the 2-matrix model*, J. Phys. A **39** (2006), no. 49, 15091–15134.
6. M. Bertola, B. Eynard, and J. Harnad, *Duality, biorthogonal polynomials and multi-matrix models*, Comm. Math. Phys. **229** (2002), no. 1, 73–120.
7. ———, *Partition functions for matrix models and isomonodromic tau functions: Random matrix theory*, J. Phys. A **36** (2003), no. 12, 3067–3083.
8. ———, *Differential systems for biorthogonal polynomials appearing in 2-matrix models and the associated Riemann–Hilbert problem*, Comm. Math. Phys. **243** (2003), no. 2, 193–240.
9. M. Bertola and B. Eynard, *The PDEs of biorthogonal polynomials arising in the two-matrix model*, Math. Phys. Anal. Geom. **9** (2006), no. 1, 23–52.
10. P. Bleher and A. Its, *Semiclassical asymptotics of orthogonal polynomials, Riemann–Hilbert problem, and universality in the matrix model*, Ann. of Math. (2) **150** (1999), no. 1, 185–266.
11. O. Bohigas, M.-J. Giannoni, and C. Schmit, *Characterization of chaotic quantum spectra and universality of level fluctuation laws*, Phys. Rev. Lett. **52** (1984), no. 1, 1–4.
12. O. Bohigas and M.-J. Giannoni, *Chaotic motion and random matrix theories*, Mathematical and Computational Methods in Nuclear Physics (Granada, 1983), Lecture Notes in Phys., vol. 209, Springer, Berlin, 1984, pp. 1–99.
13. O. Bohigas, R. U. Haq, and A. Pandey, *Higher-order correlations in spectra of complex systems*, Phys. Rev. Lett. **54** (1985), no. 15, 1645–1648.
14. É. Brézin and S. Hikami, *Characteristic polynomials of random matrices*, Comm. Math. Phys. **214** (2000), no. 1, 111–135.
15. ———, *Characteristic polynomials of real symmetric random matrices*, Comm. Math. Phys. **223** (2001), no. 2, 363–382.
16. T. A. Brody, J. Flores, J. B. French, P. A. Mello, A. Pandey, and S. S. M. Wong, *Random-matrix physics: Spectrum and strength fluctuations*, Rev. Modern Phys. **53** (1981), no. 3, 385–479.
17. W. Bryc and V. U. Pierce, *Duality of real and quaternion random matrices*, available at `arXiv:0806.3695`.
18. P. Deift, *Orthogonal polynomials and random matrices: A Riemann–Hilbert approach*, Courant Lect. Notes Math., vol. 3, New York Univ., Courant Inst. Math. Sci., New York, NY; Amer. Math. Soc., Providence, RI, 1999.
19. ———, *Universality for mathematical and physical systems*, International Congress of Mathematicians. Vol. I (Madrid, 2006), Eur. Math. Soc., Zürich, 2007, pp. 125–152.
20. P. Deift and D. Gioev, *Universality in random matrix theory for orthogonal and symplectic ensembles*, Int. Math. Res. Pap. IMRP **2** (2007), Art. ID rpm004.
21. ———, *Universality at the edge of the spectrum for unitary, orthogonal, and symplectic ensembles of random matrices*, Comm. Pure Appl. Math. **60** (2007), no. 6, 867–910.
22. P. Desrosiers, *Duality in random matrix ensembles for all β*, available at `arXiv:0801.3438`.

23. P. Di Francesco, P. Ginsparg, and J. Zinn-Justin, *2D gravity and random matrices*, Phys. Rep. **254** (1995), no. 1-2, 133.
24. F. J. Dyson, *The threefold way: Algebraic structure of symmetry groups and ensembles in quantum mechanics*, J. Mathematical Phys. **3** (1962), 1199–1215.
25. ———, *A Brownian-motion model for the eigenvalues of a random matrix*, J. Mathematical Phys. **3** (1962), 1191–1198.
26. ———, *Correlations between eigenvalues of a random matrix*, Comm. Math. Phys. **19** (1970), 235–250.
27. ———, *A class of matrix ensembles*, J. Mathematical Phys. **13** (1972), 90–97.
28. F. J. Dyson and M. L. Mehta, *Statistical theory of the energy levels of complex systems.* IV, J. Mathematical Phys. **4** (1963), 701–712.
29. K. Efetov, *Supersymmetry in disorder and chaos*, Cambridge Univ. Press, Cambridge, 1997.
30. B. Eynard, *Asymptotics of skew orthogonal polynomials*, J. Phys. A **34** (2001), no. 37, 7591–7605.
31. ———, *A concise expression for the ODE's of orthogonal polynomials*, available at `arXiv: math-ph/0109018`.
32. D. Fox and P. B. Kahn, *Higher order spacing distributions for a class of unitary ensembles*, Phys. Rev. (2) **134** (1964), B1151–B1155.
33. J. B. French, V. K. B. Kota, A. Pandey, and S. Tomsovic, *Statistical properties of many-particle spectra.* V: *Fluctuations and symmetries*, Ann. Physics **181** (1988), no. 2, 198–234; VI: *Fluctuation bounds on N-N T-noninvariance*, 235–260.
34. Y. V. Fyodorov and E. Strahov, *An exact formula for general spectral correlation function of random Hermitian matrices*, J. Phys. A **36** (2003), no. 12, 3203–3213.
35. S. Ghosh, *Non-Gaussian Ensembles of Random Matrices*, thesis (unpublished), Jawaharlal Nehru University, New Delhi, 2002.
36. ———, *Generalized Christoffel–Darboux formula for skew-orthogonal polynomials and random matrix theory*, J. Phys. A **39** (2006), no. 28, 8775–8782.
37. ———, *Skew-orthogonal polynomials, differential systems and random matrix theory*, J. Phys. A **40** (2007), no. 4, 711–740.
38. ———, *Generalized Christoffel–Darboux formula for classical skew-orthogonal polynomials*, J. Phys. A **41** (2008), no. 43, 435204.
39. ———, *Bulk asymptotics of skew-orthogonal polynomials for quartic double well potential and universality in the matrix model*, J. Math. Phys. **50** (2009), 063515.
40. S. Ghosh and A. Pandey, *Skew-orthogonal polynomials and random-matrix ensembles*, Phys. Rev. E **65** (2002), no. 4, 046221.
41. S. Ghosh, A. Pandey, S. Puri, and R. Saha, *Non-Gaussian random-matrix ensembles with banded spectra*, Phys. Rev. E **67** (2003), no. 2, 025201.
42. V. L. Girko, *Theory of random determinants*, Math. Appl. (Soviet Ser.), vol. 45, Kluwer Acad. Publ., Dordrecht, 1990.
43. T. Guhr, A. Müller-Groeling, and H. A. Weidenmüller, *Random-matrix theories in quantum physics: Common concepts*, Phys. Rep. **299** (1998), no. 4-6, 189–425.
44. F. Haake, *Quantum signatures of chaos*, 2nd ed., Springer Ser. Synergetics, Springer, Berlin, 2001.
45. R. U. Haq, A. Pandey, and O. Bohigas, *Fluctuation properties of nuclear energy levels: Do theory and experiment agree?*, Phys. Rev. Lett. **48** (1982), no. 16, 1086–1089.
46. A. A. Kapaev, *Riemann–Hilbert problem for bi-orthogonal polynomials*, J. Phys. A **36** (2003), no. 16, 4629–4640.
47. N. M. Katz and P. Sarnak, *Random matrices, Frobenius eigenvalues, and monodromy*, Amer. Math. Soc. Colloq. Publ., vol. 45, Amer. Math. Soc., Providence, RI, 1999.
48. Y. Kodama and V. U. Pierce, *Geometry of the Pfaff lattices*, Int. Math. Res. Not. IMRN **23** (2007), no. 23, Art. ID rnm120.
49. ———, *Combinatorics of dispersionless integrable systems and universality in random matrix theory*, available at `arXiv:0811.0351`.
50. A. B. J. Kuijlaars and K. T.-R. McLaughlin, *A Riemann-Hilbert problem for biorthogonal polynomials*, J. Comput. Appl. Math. **178** (2005), no. 1-2, 313–320.
51. M. L. Mehta, *A note on correlations between eigenvalues of a random matrix*, Comm. Math. Phys. **20** (1971), 245–250.

52. ______, *Matrix theory: Selected topics and useful results*, 2nd ed., Hindustan Publishing Corp., Delhi, 1989.
53. ______, *Random matrices*, 3rd ed., Pure Appl. Math. (Amst.), vol. 142, Elsevier/Academic Press, Amsterdam, 2004.
54. M. L. Mehta and G. Mahoux, *A method of integration over matrix variables.* III, Indian J. Pure Appl. Math. **22** (1991), no. 7, 531–546.
55. M. L. Mehta and J.-M. Normand, *Moments of the characteristic polynomial in the three ensembles of random matrices*, J. Phys. A **34** (2001), no. 22, 4627–4639.
56. M. L. Mehta and A. Pandey, *On some Gaussian ensembles of Hermitian matrices*, J. Phys. A **16** (1983), no. 12, 2655–2684.
57. P. A. Mello, *Macroscopic approach to universal conductance fluctuations in disordered metals*, Phys. Rev. Lett. **60** (1988), no. 11, 1089–1092.
58. P. A. Mello, P. Pereyra, and N. Kumar, *Macroscopic approach to multichannel disordered conductors*, Ann. Physics **181** (1988), no. 2, 290–317.
59. E. H. Moore, *General analysis.* I: *The algebra of matrices*, Mem. Amer. Philos. Soc., vol. 1, Amer. Philos. Soc., Philadelphia, PA, 1935.
60. M. Mulase and A. Waldron, *Duality of orthogonal and symplectic matrix integrals and quaternionic Feynman graphs*, Comm. Math. Phys. **240** (2003), no. 3, 553–586.
61. T. Nagao and P. J. Forrester, *Transitive ensembles of random matrices related to orthogonal polynomials*, Nuclear Phys. B **530** (1998), no. 3, 742–762.
62. T. Nagao and M. Wadati, *Correlation functions of random matrix ensembles related to classical orthogonal polynomials*, J. Phys. Soc. Japan **60** (1991), no. 10, 3298–3322.
63. ______, *Correlation functions of random matrix ensembles related to classical orthogonal polynomials.* II, J. Phys. Soc. Japan **61** (1992), no. 1, 78–88.
64. A. Pandey, *Statistical properties of many-particle spectra.* III: *Ergodic behavior in random-matrix ensembles*, Ann. Physics **119** (1979), no. 1, 170–191.
65. ______, *Statistical properties of many-particle spectra.* IV: *New ensembles by Stieltjes transform methods*, Ann. Physics **134** (1981), no. 1, 110–127.
66. ______, *Brownian-motion model of discrete spectra*, Chaos Solitons Fractals **5** (1995), no. 7, 1275–1285.
67. A. Pandey and S. Ghosh, *Skew-orthogonal polynomials and universality of energy-level correlations*, Phys. Rev. Lett. **87** (2001), no. 2, 024102.
68. A. Pandey and P. Shukla, *Eigenvalue correlations in the circular ensembles*, J. Phys. A **24** (1991), no. 16, 3907–3926.
69. V. U. Pierce, *A Riemann–Hilbert problem for skew-orthogonal polynomials*, J. Comput. Appl. Math. **215** (2008), no. 1, 230–241.
70. M. Plancherel and W. Rotach, *Sur les valeurs asymptotiques des polynômes d'Hermite* $H_n(x) = (-1)^n e^{x^2/2}\, d^n(e^{-x^2/2})/\, dx^n$, Comment. Math. Helv. **1** (1929), no. 1, 227–254.
71. C. E. Porter (ed.), *Statistical theories of spectra: Fluctuations*, Perspect. Phys., vol. 1, Academic Press, New York, NY, 1965.
72. M. K. Şener and J. J. M. Verbaarschot, *Universality in chiral random matrix theory at* $\beta = 1$ *and* $\beta = 4$, Phys. Rev. Lett. **81** (1998), no. 2, 248–251.
73. H.-J. Stöckmann, *Quantum chaos: An introduction*, Cambridge Univ. Press, Cambridge, 1999.
74. A. Stojanovic, *Universality in orthogonal and symplectic invariant matrix models with quartic potential*, Math. Phys. Anal. Geom. **3** (2000), no. 4, 339–373.
75. ______, *Une approche par les polynômes orthogonaux pour des classes de matrices aléatoires orthogonalement et symplectiquement invariantes: Application à l'universalité de la statistique locale des valeurs propres*, Technical Report 00-01-006, Research Center BiBoS, available at `www.physik.uni-bielefeld.de/bibos/`.
76. ______, *Errata: "Universality in orthogonal and symplectic invariant matrix models with quartic potential"*, Math. Phys. Anal. Geom. **7** (2004), no. 4, 347–349.
77. G. Szegő, *Orthogonal polynomials*, Amer. Math. Soc. Colloq. Publ., vol. 23, Amer. Math. Soc., New York, NY, 1939.
78. H. Widom, *On the relation between orthogonal, symplectic and unitary matrix ensembles*, J. Statist. Phys. **94** (1999), no. 3-4, 347–363.